T0331216

Ultra-Wideband Communications Systems

1807 ⊛WILEY 2007 BICENTENNIAL

THE WILEY BICENTENNIAL—KNOWLEDGE FOR GENERATIONS

*E*ach generation has its unique needs and aspirations. When Charles Wiley first opened his small printing shop in lower Manhattan in 1807, it was a generation of boundless potential searching for an identity. And we were there, helping to define a new American literary tradition. Over half a century later, in the midst of the Second Industrial Revolution, it was a generation focused on building the future. Once again, we were there, supplying the critical scientific, technical, and engineering knowledge that helped frame the world. Throughout the 20th Century, and into the new millennium, nations began to reach out beyond their own borders and a new international community was born. Wiley was there, expanding its operations around the world to enable a global exchange of ideas, opinions, and know-how.

For 200 years, Wiley has been an integral part of each generation's journey, enabling the flow of information and understanding necessary to meet their needs and fulfill their aspirations. Today, bold new technologies are changing the way we live and learn. Wiley will be there, providing you the must-have knowledge you need to imagine new worlds, new possibilities, and new opportunities.

Generations come and go, but you can always count on Wiley to provide you the knowledge you need, when and where you need it!

WILLIAM J. PESCE
PRESIDENT AND CHIEF EXECUTIVE OFFICER

PETER BOOTH WILEY
CHAIRMAN OF THE BOARD

Ultra-Wideband Communications Systems

Multiband OFDM Approach

W. Pam Siriwongpairat
Meteor Communications Corporation

K. J. Ray Liu
University of Maryland

IEEE PRESS

BICENTENNIAL
1807
WILEY
2007
BICENTENNIAL

A JOHN WILEY & SONS, INC., PUBLICATION

Library of Congress Cataloging-in-Publication Data:

Siriwongpairat, W. Pam, 1978–
   Ultra-wideband communication systems : multiband OFDM approach / W. Pam
Siriwongpairat. K.J. Ray Liu.
      p. cm.
   ISBN 978-0-470-07469-5 (cloth)
   1. Orthogonal frequency division multiplexing.   2. Ultra-wideband devices.
I. Liu, K. J. Ray, 1961–   II. Title.
TK5103.484.S57 2007
621.384—dc22

                                                        2007012087

Printed in the United States of America.
10 9 8 7 6 5 4 3 2 1

To my parents, Sawai and Pimjai Siriwongpairat

—WPS

To Jeffry and Joanne Liu

—KJRL

CONTENTS

PREFACE

Ultra-wideband (UWB) has emerged as a technology that offers great promise to satisfy the growing demand for low-cost, high-speed digital wireless indoor and home networks. The enormous bandwidth available, the potential for high data rates, and the potential for small size and low processing power along with low implementation cost all present a unique opportunity for UWB to become a widely adopted radio solution for future wireless home-networking technology.

UWB is defined as any transmission that occupies a bandwidth of more than 20% of its center frequency, or more than 500 MHz. In 2002, the Federal Communications Commission (FCC) has mandated that UWB radio transmission can legally operate in the range 3.1 to 10.6 GHz at a transmitter power of -41.3 dBm/MHz. The use of UWB technology under the FCC guidelines can provide enormous capacity over short ranges. This can be seen by considering Shannon's capacity equation, which shows that increasing channel capacity requires linear increases in bandwidth, whereas similar channel capacity increases would require exponential increases in power. Currently, UWB technology is able to support various data rates, ranging from 55 to 480 Mbps, over distances up to 10 meters. In addition, it is expected that UWB devices will consume very little power and silicon area, as well as provide low-cost solutions that can satisfy consumer market demands.

Nevertheless, to fulfill these expectations, UWB research and development have to cope with several challenges, including high-sensitivity synchronization, ability to capture the multipath energy, low-complexity constraints, strict power limitations, scalability, and flexibility. Such challenges require advanced digital signal-processing expertise to develop systems that could take advantage of the UWB spectrum and support future indoor wireless applications.

This book provides comprehensive coverage of the fundamental issues in UWB technology, with particular focus on the *multiband orthogonal frequency-division multiplexing* (multiband OFDM) approach, which has been a leading proposal in the IEEE 802.15.3a standard and has recently been adopted in the ECMA standard for wireless personal area networks. The book also explores several major advanced state-of-the-art technologies to enhance the performance of the standardized multiband OFDM approach.

In Chapter 1 we provide an introduction to UWB communications. In this chapter we present a comprehensive overview of UWB radios and review the historical development of UWB. Then we discuss the advantages, challenges, and applications of UWB technology.

In Chapter 2 we describe the characteristics of UWB channels and establish a mathematical channel model for the analysis in subsequent chapters.

Chapter 3 provides an overview of UWB single-band approaches. In this chapter we describe the signal modeling and transceiver design of single-band approaches. Then we derive the bit-error-rate performance of single-band UWB systems. Performance analysis is provided for both single- and multiple-antenna UWB systems.

Chapter 4 is an overview of the multiband OFDM approach. In this chapter we provide the fundamental background for the multiband OFDM approach used in subsequent chapters.

In Chapter 5 we extend the multiband OFDM approach to a multiple-antenna system. In this chapter we first describe a multiple-input multiple-output coding framework for UWB multiband OFDM systems. We show that a combination of space–time–frequency coding and hopping multiband OFDM modulation can fully exploit all of the available spatial and frequency diversities inherent in UWB environments.

In Chapter 6 we analyze the performance of UWB multiband OFDM systems under realistic UWB channels. We characterize pairwise error and outage probabilities in UWB multiband OFDM systems, based on the multipath-clustering phenomenon of UWB channels. The analysis is first provided for single-antenna systems, then extended to multiantenna systems.

Chapter 7 extends the performance analysis in Chapter 6 to a more practical scenario. Specifically, we provide a performance analysis of multiband OFDM systems that not only captures the characteristics of realistic UWB channels, but also takes into consideration the imperfection of frequency and timing synchronization and the effect of intersymbol interference.

In Chapter 8 we introduce a differential UWB scheme as an alternative approach that bypasses channel estimation and provides a good trade-off between performance and complexity in UWB communications systems. We review a basic concept of differential OFDM, then describe a differential multiband OFDM system and analyze its performance. Both single- and multiantenna differential UWB systems are considered.

In Chapter 9 we present a power-controlled channel allocation scheme for multiband OFDM systems. The scheme allocates subbands and transmitted power among UWB users to minimize overall power consumption. This allows a UWB multiband OFDM system to operate at a low transmitter power level while still achieving the performance desired.

In Chapter 10 we introduce cooperative communications in UWB systems to enhance the performance and coverage of UWB by exploiting the broadcasting nature of wireless channels and cooperation among UWB devices. The principal concept of cooperative communications is presented, and then it is applied to multiband OFDM systems. Performance analysis and optimum power allocation of cooperative UWB multiband OFDM systems are provided.

We would like to express our gratitude to Dr. Thanongsak Himsoon, who contributed the major part of Chapter 8, and to Hung-Quoc Lai for his contributions to Chapters 4 and 7. We also would like to thank Dr. Weifeng Su, Dr. Zhu Han, and Dr. Masoud Olfat for their research represented in several works described in the book.

W. PAM SIRIWONGPAIRAT
K. J. RAY LIU

Meteor Communications Corporation
Kent, Washington
University of Maryland
College Park, Maryland

1

INTRODUCTION

In the near future, indoor communications of any digital data—from high-speed signals carrying multiple HDTV programs to low-speed signals used for timing purposes—will be shared over a digital wireless network. Such indoor and home networking is unique, in that it simultaneously requires high data rates (for multiple streams of digital video), very low cost (for broad consumer adoption), and very low power consumption (for embedding into battery-powered handheld appliances). With its enormous bandwidth, *ultra-wideband* (UWB) provides a promising solution to satisfying these requirements and becomes an attractive candidate for future wireless indoor networks.

We begin with an overview of UWB radios and review the historical development of UWB. Next, we present the key benefits of UWB. Then we discuss the application potential of UWB technology for wireless communications. Finally, an overview of UWB transmission schemes is presented, and the challenges in designing UWB communication systems are discussed.

1.1 OVERVIEW OF UWB

The concept of UWB was developed in the early 1960s through research in time-domain electromagnetics, where impulse measurement techniques were used to characterize the transient behavior of a certain class of microwave networks [Ros63]. In the late 1960s, impulse measurement techniques were applied to the design of wideband antenna elements, leading to the development of short-pulse radar and communications systems. In 1973, the first UWB communications patent was awarded for a short-pulse receiver [Ros73]. Through the late 1980s, UWB was referred to as *baseband, carrier-free*, or *impulse* technology. The term *ultra-wideband* was coined in approximately 1989 by the U.S. Department of Defense. By 1989, UWB theory, techniques, and many implementation approaches had been developed for a wide range of applications, such as radar, communications, automobile collision avoidance,

Ultra-Wideband Communications Systems: Multiband OFDM Approach, By W. Pam Siriwongpairat and K. J. Ray Liu
Copyright © 2008 John Wiley & Sons, Inc.

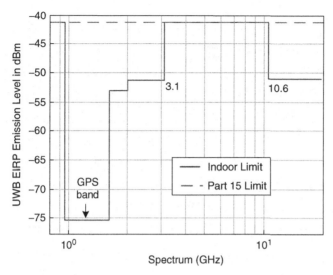

Figure 1.1 UWB spectral mask for indoor communication systems.

positioning systems, liquid-level sensing, and altimetry. However, much of the early work in the UWB field occurred in the military or was funded by the U.S. government within classified programs. By the late 1990s, UWB technology had become more commercialized and its development had accelerated greatly. For an interesting and informative review of UWB history, the interested reader is referred to [Bar00].

A substantial change in UWB history occurred in February 2002, when the, U.S. Federal Communications Commission (FCC) issued UWB rulings that provided the first radiation limitations for UWB transmission and permitted the operation of UWB devices on an unlicensed basis [FCC02]. According to the FCC rulings, UWB is defined as any transmission scheme that occupies a fractional bandwidth greater than 0.2 or a signal bandwidth of more than 500 MHz. The fractional bandwidth is defined as B/f_c, where $B \triangleq f_H - f_L$ represents the -10 dB bandwidth and $f_c \triangleq (f_H + f_L)/2$ denotes the center frequency. Here f_H and f_L are the upper frequency and the lower frequency, respectively, measured at -10 dB below the peak emission point. Based on [FCC02], UWB systems with $f_c > 2.5$ GHz need to have a -10 dB bandwidth of at least 500 MHz, whereas UWB systems with $f_c < 2.5$ GHz need to have a fractional bandwidth of at least 0.2. The FCC has mandated that UWB radio transmission can legally operate in the range 3.1 to 10.6 GHz, with the power spectral density (PSD) satisfying a specific spectral mask assigned by the FCC. In particular, Fig. 1.1 illustrates the UWB spectral mask for indoor communications under Part 15 of the FCC's rules [FCC02]. According to the spectral mask, the PSD of a UWB signal measured in the 1-MHz bandwidth must not exceed -41.3 dBm, which complies with the Part 15 general emission limits for successful control of radio interference. For particularly sensitive bands such as the global positioning system (GPS) band (0.96 to 1.61 GHz), the PSD limit is much lower. As depicted in

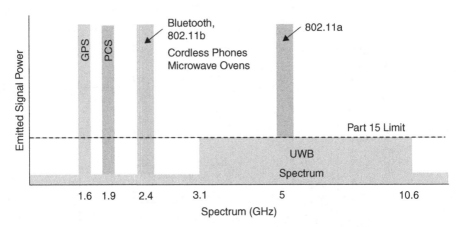

Figure 1.2 Spectrum of UWB and existing narrowband systems.

Fig. 1.2, such a ruling allows UWB devices to overlay existing narrowband systems while ensuring sufficient attenuation to limit adjacent channel interference. Although at present, UWB operation is permitted only in the United States, regulatory efforts are under way in many countries, especially in Europe and Japan [Por03]. Market drivers for UWB technology are many, even at this early stage, and are expected to include new applications in the next few years.

1.2 ADVANTAGES OF UWB

Due to its ultra-wideband nature, UWB radios come with unique benefits that are attractive for radar and communications applications. The principal advantages of UWB can be summarized as follows [Kai05]:

- Potential for high data rates
- Extensive multipath diversity
- Potential small size and processing power together with low equipment cost
- High-precision ranging and localization at the centimeter level

The extremely large bandwidth occupied by UWB gives the potential of very high theoretical capacity, yielding very high data rates. This can be seen by considering *Shannon's capacity equation* [Pro01],

$$C = B \log \left(1 + \frac{S}{N} \right), \tag{1.1}$$

where C is the maximum channel capacity, B the signal bandwidth, S the signal power, and N the noise power. Shannon's equation shows that the capacity can be improved by increasing the signal bandwidth or the signal power. Moreover, it shows that increasing channel capacity requires linear increases in bandwidth, while similar channel capacity increases would require exponential increases in power. Thus, from

Shannon's equation we can see that UWB system has a great potential for high-speed wireless communications.

Conveying information with ultrashort-duration waveforms, UWB signals have low susceptibility to multipath interference. Multipath interference occurs when a modulated signal arrives at a receiver from different paths. Combining signals at the receiver can result in distortion of the signal received. The ultrashort duration of UWB waveforms gives rise to a fine resolution of reflected pulses at the receiver. As a result, UWB transmissions can resolve many paths, and are thus rich in multipath diversity.

The low complexity and low cost of UWB systems arises from the carrier-free nature of the signal transmission. Specifically, due to its ultrawide bandwidth, the UWB signal may span a frequency commonly used as a carrier frequency. This eliminates the need for an additional radio-frequency (RF) mixing stage as required in conventional radio technology. Such an omission of up/down-conversion processes and RF components allows the entire UWB transceiver to be integrated with a single CMOS implementation. Single-chip CMOS integration of a UWB transceiver contributes directly to low cost, small size, and low power.

The ultrashort duration of UWB waveforms gives rise to the potential ability to provide high-precision ranging and localization. Together with good material penetration properties, UWB signals offer opportunities for short-range radar applications such as rescue and anticrime operations as well as in surveying and in the mining industry.

1.3 UWB APPLICATIONS

UWB technology can enable a wide variety of applications in wireless communications, networking, radar imaging, and localization systems. For wireless communications the use of UWB technology under the FCC guidelines [FCC02] offers significant potential for the deployment of two basic communications systems:

- High-data-rate short-range communications: high-data-rate wireless personal area networks
- Low-data-rate and location tracking: sensor, positioning, and identification networks

The high-data-rate WPANs can be defined as networks with a medium density of active devices per room (5 to 10) transmitting at data rates ranging from 100 to 500 Mbps within a distance of 20 m. The ultrawide bandwidth of UWB enables various WPAN applications, such as high-speed wireless universal serial bus (WUSB) connectivity for personal computers (PCs) and PC peripherals, high-quality real-time video and audio transmission, file exchange among storage systems, and cable replacement for home entertainment systems.

Recently, the IEEE 802.15.3 standard task group has established the 802.15.3a study group [TG3a] to define a new physical layer concept for high-data-rate WPAN applications. A major goal of this study group is to standardize UWB wireless radios for indoor WPAN transmissions. The goal for the IEEE 802.15.3a standard is to provide a higher-speed physical layer for the existing approved 802.15.3 standard for applications that involve imaging and multimedia. The work of the 802.15.3a study group includes standardizing the channel model to be used for UWB system evaluation.

Alternatively, UWB transmission can trade a reduction in data rate for an increase in transmission range. Under the low-rate operation mode, UWB technology could be beneficial and potentially useful in sensor, positioning, and identification networks. A sensor network comprises a large number of nodes spread over a geographical area to be monitored. Depending on the specific application, the sensor nodes can be static or mobile. Key requirements for sensor networks operating in challenging environments include low cost, low powers, and multifunctionality. With its unique properties of low complexity, low cost, and low power, UWB technology is well suited to sensor network applications [Opp04]. Moreover, due to the fine time resolution of UWB signals UWB-based sensing has the potential to improve the resolution of conventional proximity and motion sensors. The low-rate transmission, combined with accurate location tracking capabilities, offers an operational mode known as *low-data-rate and location tracking*.

The IEEE also established the 802.15.4 study group to define a new physical layer concept for low-data-rate applications utilizing UWB technology at the air interface. The study group addressed new applications which require only moderate data throughput but long battery life, such as low-rate wireless personal area networks, sensors, and small networks.

1.4 UWB TRANSMISSION SCHEMES

Although the FCC has regulated the spectrum and transmitter power levels for a UWB, there is currently no standard for a UWB transmission scheme. Various pulse generation techniques have been proposed to use the 7.5-GHz license-free UWB spectrum. Generally, UWB transmission approaches can be categorized into two main approaches: single-band and multiband. Figure 1.3 illustrates UWB signals in the time and frequency domains when single and multiband approaches are employed.

A traditional UWB technology is based on single-band systems employing carrier-free or impulse radio communications [Sch93, Win98, Wel01, Foe02a, Rob03, Bou03]. *Impulse radio* refers to the generation of a series of impulselike waveforms, each of duration in the order of hundreds of picoseconds. Each pulse occupies a bandwidth of several gigahertz that must adhere to the spectral mask requirements. The information is modulated directly into the sequence of pulses. Typically, one pulse carries the information for 1 bit. Data could be modulated using either pulse amplitude modulation (PAM) or pulse position modulation (PPM). Multiple users can be supported using the time-hopping or direct-sequence spreading approaches. This type of transmission does not require the use of additional carrier modulation, as the pulse will propagate well in the radio channel. The technique is therefore a baseband

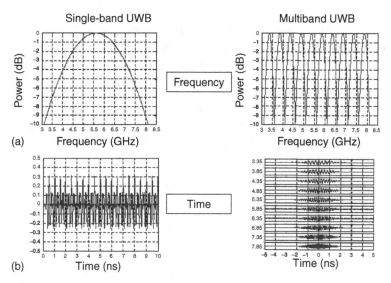

Figure 1.3 UWB transmission approaches: (a) single- and (b) multiband approaches [Dis03].

signal approach. However, the single-band system faces a challenging problem in building RF and analog circuits and in designing a low-complexity receiver that can capture sufficient multipath energy.

To overcome the drawback of single-band approaches, multiband approaches were proposed in [Sab03, Foe03a, Bat03, Bat04]. Instead of using the entire UWB frequency band to transmit information, the multiband technique divides the UWB frequency band from 3.1 to 10.6 GHz into several smaller bands, referred to as *subbands*. Each subband occupies a bandwidth of at least 500 MHz, in compliance with FCC regulations [FCC02]. By interleaving the transmitted symbols across subbands, multiband approaches can maintain the power being transmitted as if a large GHz bandwidth were being utilized. The advantage is that multiband approaches allow information to be processed over a much smaller bandwidth, thereby reducing overall design complexity as well as improving spectral flexibility and worldwide compliance.

Recently, a multiband OFDM approach that utilizes a combination multiband approach and orthogonal frequency-division multiplexing (OFDM) technique was proposed [Bat03]. The OFDM technique is efficient at collecting multipath energy in highly dispersive channels, as is the case for most UWB channels [Bat04]. Moreover, OFDM allows each subband to be divided into a set of orthogonal narrowband channels (with a much longer symbol period duration). The major difference between multiband and traditional OFDM schemes is that multiband OFDM symbols are not sent continually on a single frequency band; instead, they are interleaved over different subbands across both time and frequency. Multiple access to the multiband approach is enabled by the use of suitably designed frequency-hopping sequences over the set of subbands. A frequency synthesizer can be utilized to perform frequency hopping. By using proper time–frequency codes, a multiband system provides both frequency diversity and multiple access capability [Bat04]. The multiband OFDM approach has been a leading proposal for the IEEE 802.15.3a WPAN standard [TG3a] and has been approved as the UWB standard by the European Computer Manufacturers Association (ECMA) [ECM05].

There are many trade-offs in the UWB approaches described above. The single-band approach benefits from a coding gain achieved through the use of time-hopping or direct-sequence spreading, exploits Shannon's principles to a greater degree than does the multiband approach, has greater precision for position location, and realizes better spectrum efficiency. However, it has less flexibility with regard to foreign spectral regulation and may be too broadband if foreign governments choose to limit their UWB spectral allocations to smaller ranges than authorized by the FCC. On the other hand, the multiband approach has is its main advantage the ability for finer-grained control of the transmitter PSD so as to maximize the average power transmitted while meeting the spectral mask. It allows for peaceful coexistence with flexible spectral coverage and is easier to adopt to different worldwide regulatory environments. Moreover, processing over a smaller bandwidth eases the requirement on analog-to-digital converter sampling rates and, consequently, facilitates greater digital processing. Furthermore, in the UWB multiband OFDM approach, due to the increased length of the OFDM symbol period, the modulation method can successfully reduce the effects of intersymbol interference (ISI). Nevertheless, this robust multipath tolerance comes at the price of increased transceiver complexity, the need to combat intercarrier interference (ICI), and tighter linear constraint on amplifying circuit elements.

1.5 CHALLENGES FOR UWB

Although UWB technology has several attractive properties that make it a promising technology for future short-range wireless communications and many other applications, some challenges must be overcome to fulfill these expectations.

The transmitter power level of UWB signals is strictly limited in order for UWB devices to coexist peacefully with other wireless systems. Such strict power limitation poses significant challenges when designing UWB systems. One major challenge is to achieve the performance desired at an adequate transmission range using limited transmitter power. Another challenge is to design UWB waveforms that efficiently utilize the bandwidth and power allowed by the FCC spectral mask. Moreover, to ensure that the transmitter power level satisfies the spectral mask, adequate characterization and optimization of transmission techniques (e.g., adaptive power control, duty cycle optimization) may be required.

The ultrashort duration of UWB pulses leads to a large number of resolvable multipath components at the receiver. In particular, the UWB signal received contains many delayed and scaled replicas of the pulses transmitted. Additionally, each resolvable pulse undergoes different channel fading, which makes multipath energy capture a challenging problem in UWB system design. For example, if a RAKE receiver [Proa1] is used to collect the multipath energy, a large number of fingers is needed to achieve the performance desired.

Design challenges also exist in the areas of modulation and coding techniques that are suitable for UWB systems. Originally, UWB radio was used for military applications, where multiuser transmission and achieving high multiuser capacity are not major concerns. However, these issues become very important in commercial applications, such as high-speed wireless home networks. Effective coding and modulation schemes are thus necessary to improve UWB multiuser capacity as well as system performance.

One design challenge is the impact of narrowband interference on UWB receivers. Specifically, the UWB frequency band overlaps with that of IEEE 802.11a wireless local area networks (WLANs). The signals from 802.11a devices represent in-band interference for the UWB receiver front end.

Other design challenges include scalable system architectures and spectrum flexibility. UWB potential applications include both high-rate applications (e.g., images and video) and lower-rate applications (e.g., computer peripheral support). Thus, the UWB transceiver must be able to support a wide range of data rates. Furthermore, the unlicensed nature of the UWB spectrum makes it essential for UWB devices to coexist with devices that share the same spectrum. However, it is challenging to design UWB systems with spectrum flexibility that allows UWB devices to coexist effectively with other wireless technologies and to meet potentially different regulatory requirements in different regions of the world.

2

CHANNEL CHARACTERISTICS

Analysis and design of UWB communication systems require an accurate channel model to determine the that can be achieved, to design efficient modulation and coding schemes, and to develop associated signal-processing algorithms. Although narrowband wireless channels have been well documented in the literature (see, e.g., [Ber00, Bla99, Dur03] and references therein), they cannot be generalized directly to UWB channels. In particular, the narrowband channels were constructed based on a signal bandwidth of less than 20 MHz. The radiation in UWB systems, on the other hand, can cover as much as 10 GHz of bandwidth. Such a large bandwidth gives rise to important differences between UWB and narrowband channels, especially with respect to the number of resolvable paths and arrival times of multipath components [Foe03b]. In particular, the large bandwidth of a UWB waveform considerably increases the ability of a receiver to resolve a variety of reflections in UWB channels. As a result, the signal received contains a significant number of resolvable multipath components. Additionally, due to the fine time resolution of a UWB waveform, the multipath components tend to occur in a cluster rather than in a continuum as is common for narrowband channels.

In recent years, a lot of research effort has been devoted to UWB channel modeling in order to construct channel models that are able to capture these important characteristics of UWB channels. In [Win02], simulation results for indoor communications using UWB signals were presented; the UWB channel is modeled as a tap-delay-line fading model, and the amplitude of the multipath coefficients is characterized by Nakagami-m distribution. A time-domain measurement campaign has been conducted in [Win98], and the measurement results were later used to develop UWB channel models in [Cra02]. The time-domain measurement approach generally excites the channel by a short pulse and has samples of the channel response recorded at the receiver. UWB channel modeling using a frequency-domain measurement approach was presented in [Gha02, Str01, Kun02]. In the frequency-domain measurement approach, a vector network analyzer is used to record the channel frequency response (instead of the time-domain response); measurements in

Ultra-Wideband Communications Systems: Multiband OFDM Approach, By W. Pam Siriwongpairat and K. J. Ray Liu
Copyright © 2008 John Wiley & Sons, Inc.

the frequency domain are then converted to the time domain using inverse Fourier transform.

To have a common channel model for the evaluation of UWB communications systems, standardization groups SG3a and SG4a recently established within the IEEE 802.15 have been working to set up standard models of the UWB channel. A UWB channel model for the IEEE 802.15.3a standard [TG3a] has been established in [Foe03b], while the model for the IEEE 802.15.4a [TG4a] is currently in progress.

In this chapter we provide an overview of UWB channel models as offered in the recent literature. The propagation models are divided into two categories: large- and small-scale models. The large-scale models characterize signal power over long transmitter–receiver separation distances and the small-scale models characterize signal behavior over a very short distance (up to 30 m outdoors or a few meters in indoor scenarios). At the end of the chapter we present large- and small-scale models of the standard UWB channel model that has been established by the IEEE 802.15.3a standards task group.

2.1 LARGE-SCALE MODELS

Large-scale models are described by path loss models and shadowing [Stu00]. Specifically, path loss models describe signal attenuation between a transmitter and a receiver. The path loss models provide an average value of the signal power as a function of the propagation distance. Shadowing characterizes the slow variation of the signal envelope over time around the deterministic path loss value. The shadowing behavior is caused by large obstructions (e.g., buildings or landscape) that are distant from the receiver.

Both path loss and shadowing are affected by several factors, including local terrain characteristics, signal bandwidth, and carrier frequency. To date, most of existing path loss and shadowing models for UWB communications are based on narrowband models, with some modifications to account for the fact that the UWB bandwidth can cover a range of several gigahertz. The UWB path loss and shadowing models are described in the following subsections.

2.1.1 Path Loss Models

Path loss is defined as the ratio between the signal power at the transmitter and the signal power at the receiver. It is affected by the distance between the transmitter and the receiver, the signal bandwidth, the carrier frequency, the antenna heights, and the terrain characteristics (e.g., buildings and hills).

In an ideal situation where the electromagnetic field propagates in the absence of any reflections, the path loss—referred to as free-space path loss—is modeled as [Stu00]

$$\overline{PL}_f(d) = \left(\frac{4\pi f_c d}{c}\right)^2, \tag{2.1}$$

where d represents the distance between a transmitter and a receiver, c is the speed of light, and f_c is the center frequency. For narrowband communications, the center frequency is given by $f_c = (f_L + f_H)/2$, where f_L and f_H are the lower and upper frequencies, respectively.

In practice, the path loss is affected not only by the free-space loss but also by the signal refraction, diffraction, and scattering. A more general path loss model that is able to capture the essence of signal propagation without resorting to complicated path loss models is given by [Stu00]

$$\overline{PL}(d) = \overline{PL}(d_0) \left(\frac{d}{d_0} \right)^{\kappa},$$

(2.2)

where d_0 is the reference distance (typically, 1 m), $\overline{PL}(d_0)$ is the measured path loss at the reference distance, and κ is the path loss exponent (can be obtained through measurements). Typically, $\kappa = 2$ in free space, $\kappa = 3.5$ in a rural area, $\kappa = 4$ in a suburban area, and $\kappa = 4.5$ in a dense urban area. Due to scattering phenomena in the antenna near field, this model is generally valid only at a transmission distance $d > d_0$.

In UWB communications, the frequency dependence of the propagation parameters can be significant. Such frequency dependence can be taken into account in the UWB path loss model as [Kun02, Alv03]

$$\overline{PL}(f, d) = \overline{PL_f}(f)\overline{PL_d}(d),$$

(2.3)

where $\overline{PL_f}(f)$ denotes the frequency-dependent path loss and $\overline{PL_d}(d)$ denotes the distance-dependent path loss. In [Kun02] the frequency-dependent path loss is modeled as

$$\sqrt{\overline{PL_f}(f)} \propto f^{-m}, \quad m \in [0.8, 1.4],$$

(2.4)

whereas in [Alv03] $\overline{PL_f}(f)$ is modeled as

$$\log_{10}(\overline{PL_f}(f)) \propto \exp(-\delta f), \quad \delta \in [1.0, 1.4].$$

(2.5)

2.1.2 Shadowing

Shadowing is defined as the slow variation in signal attenuation around the mean path loss value. Several measurements have shown that the shadowing behavior in UWB communications is quite similar to that in narrowband systems. Specifically, empirical studies have shown that the shadowing effect in UWB channels is lognormally distributed (i.e., the signal level measured (in decibels)) at a specific transmitter–receiver separation has a normal distribution with mean 0 dB and variance σ^2.

For a fixed distance d, the combined effect of path loss and shadowing can be modeled as [Stu00]

$$PL(d) = \overline{PL}(d)X_{\sigma},$$

(2.6)

where $\overline{PL}(d)$ is the mean path loss and X_σ denotes the shadowing effect. From the path loss model in equations (2.3), (2.6)can be expressed as

$$PL(d)[\text{dB}] = \overline{PL}(d_0)[\text{dB}] + \kappa \cdot 10 \log_{10} \frac{d}{d_0} + X_\sigma[\text{dB}], \tag{2.7}$$

where $X_\sigma[\text{dB}]$ follows zero-mean normal distribution. The standard deviation σ of X_σ [dB] depends on the specific measurement environment and antenna heights.

2.2 SMALL-SCALE MODELS

A small-scale model or fading is used to describe a rapid fluctuation of the amplitudes, phases, or multipath delays in the signal received over a short period of time. Small-scale fading is due to constructive and destructive interference of the multipath components, which arrive at the receiver at slightly different times. In contrast to the shadowing, small-scale fading is caused by the effects of objects that are close to the receiver.

For a narrowband signal with a bandwidth less than the coherence bandwidth of the propagation channel, multipath components arrive continuously and severe multipath fading can be observed. When a large number of multipath components are observed at the receiver within its resolution time, the central limit theorem is commonly invoked to model the amplitude of the signal received as Rayleigh distributed. Rayleigh fading is therefore used extensively for channel models in many narrowband systems.

In UWB systems, on the other hand, the ultralarge bandwidth of UWB signals significantly increases the ability of the receiver to resolve the multipath components. This characteristic of UWB systems can give rise to two major effects that make UWB channels different from that of narrowband systems. These two effects are as follows:

1. The number of multipath components that arrive at the receiver within the period of an ultrashort waveform is much smaller as the duration becomes shorter. Consequently, the channel fading is not as severe as that in narrowband channels, and Rayleigh fading may not perfectly match the amplitude of the signal received. In addition, a large number of resolvable multipath components can be observed at the receiver.

2. Since the multipath components can be resolved on a very fine time duration, the time of arrival of the multipath components may not be continuous. In other words, there are empty delay bins (bins containing no energy) between the arriving multipath components. In UWB systems, the channel measurements showed multipath arrivals in clusters rather than in a continuum as is common for narrowband channels. In particular, due to the very fine resolution of UWB

waveforms, different objects or walls in a room could contribute different clusters of multipath components.

A reliable channel model that captures such important characteristics of UWB channel is therefore critical for the analysis and design of UWB systems. Recently, the IEEE 802.15.3a standards task group formed a subgroup to establish a common channel model for UWB systems. The three main indoor channel models considered in the standard are described below.

2.2.1 Tap-Delay-Line Fading Model

A simple model for characterization of a UWB channel is the tap-delay-line fading model [Win02, Foe03b], in which the signal received is a sum of the replicas of the signal transmitted, being related to the reflecting, scattering, and/or deflecting objects via which the signal propagates. Such a tap-delay-line fading model allows frequency selectivity of UWB channels to be taken into consideration. Under the tap-delay-line fading model, the channel impulse response can be described as [Pro01]

$$h(t) = \sum_{l=0}^{L-1} \alpha(l)\delta(t - \tau_l), \tag{2.8}$$

where $\alpha(l)$ is the multipath gain coefficient of the lth path, L denotes the number of resolvable multipath components, and τ_l represents the path delay of the lth path. In conventional narrowband systems, it is well established that the amplitude of the lth path, $|\alpha(l)|$, is modeled as a Rayleigh random variable with a probability density function (PDF) [Pro1]

$$p_{|\alpha(l)|}(x) = \frac{x}{\Omega_l} \exp\left(-\frac{x^2}{\Omega_l}\right), \tag{2.9}$$

where $\Omega_l = E[|\alpha(l)|^2]$ denotes the average power of the lth path. Here $p_X(x)$ represents the PDF of x, and $E[\cdot]$ stands for the expectation operation. In UWB systems the number of components falling within each delay bin is much smaller, which leads to a change in statistics. Various alternative distributions have been used in the literature.

- *Lognormal distribution.* This has been suggested in [Foe02b, Gha02]. The lognormal distribution is given by [Pro1]

$$p_{|\alpha(l)|}(x) = \frac{20/\ln 10}{x\sqrt{2\pi\Omega_l}} \exp\left(\frac{(10\log_{10}(x^2) - \mu_l)^2}{2\Omega_l}\right), \tag{2.10}$$

where Ω_l is the variance of the local mean $|\alpha(l)|$ and μ_l is the mean of $|\alpha(l)|$ in decibels. The lognormal distribution has the advantage that the fading statistics of the small-scale statistics and the large-scale variations have the same form; the superposition of lognormal variables can also be well approximated by a lognormal distribution. The drawback is that it is difficult to obtain insightful performance analysis, especially for a MIMO system, under a lognormal fading model.

- *Nakagami distribution.* It has been suggested in [Win02] that the amplitude of a multipath coefficient can be modeled by Nakagami-m distribution [Nak60]:

$$p_{|\alpha(l)|}(x) = \frac{2}{\Gamma(m)} \left(\frac{m}{\Omega_l} \right)^m x^{2m-1} \exp\left(-\frac{m}{\Omega_l} x^2 \right), \qquad (2.11)$$

where $\gamma(\cdot)$ denotes the gamma function, $m \geq 1/2$ is the Nakagami fading parameter, and Ω_l is the mean-square value of the fading amplitude. The fading parameter, m, describes the severity of the fading. The smaller the m, the more severe the fading, with $m = 1$ and $m = \infty$ corresponding to Rayleigh fading and nonfading channels, respectively. The advantage of Nakagami-m statistics is that they can model a wide range of fading conditions by adjusting their fading parameters. In fact, Nakagami-m distributions with large values of m are similar to lognormal distributions. Furthermore, if the amplitude is Nakagami distributed, the power is gamma distributed, which enables closed-form performance formulation.

Although the tap-delay-line fading model is able to capture frequency selectivity, it does not reflect the clustering characteristic of UWB channels. To capture the clustering property, an approach that models multipath arrival times using a statistically random process based on the Poisson process has been considered. Specifically, the multipath arrival times τ_l can be characterized by a Poisson process with a constant arrival rate λ as

$$P_r(\tau_l - \tau_{l-1} > t) = \exp(-\lambda t). \qquad (2.12)$$

In other words, the interarrival time of multipath components is exponentially distributed; that is, given a certain arrival time τ_{l-1} for the previous time, the PDF for the arrival of path l can be written as

$$p_{\tau_l}(\tau_l \mid \tau_{l-1}) = \lambda \exp[-\lambda(\tau_l - \tau_{l-1})], \qquad l > 0. \qquad (2.13)$$

Two mathematical models that reflect this clustering are the Δ–K model [Has93] and the Saleh–Valenzuela (S-V) model [Sal87].

2.2.2 Δ–K Model

The Δ–K model was introduced for the outdoor environment, and popularized for the indoor scenario by [Has93]. The Δ–K model defines two states: state A, where the arrival rate of paths is λ, and state B, where the rate is $K\lambda$. The model starts in state A. If a path arrives at time t, a transition is made to state B for a minimum of time λ. If no path arrives during that time, the model reverts to state A; otherwise, it remains in state B. The Δ–K model was used for UWB channels in [Pen02, Zhu02].

Note that both the number of clusters and the duration of the clusters become random variables in this model. Specifically, the number of clusters is a random variable whose realization is determined by how often the system enters state B. In addition, the clusters are strictly separated from each other, the duration between two clusters is determined by the parameter Δ.

2.2.3 Saleh–Valenzuela Model

The Saleh–Valenzuela (S-V) model [Sal87] was introduced for a wideband indoor channel. In the S-V model multipath arrivals are grouped into two different categories: a cluster arrival and a ray arrival within a cluster. This model requires four main parameters: the cluster arrival rate, the ray arrival rate within a cluster, the cluster decay factor, and the ray decay factor. The channel impulse response of the S-V model is modeled by

$$h(t) = \sum_{c=0}^{C}\sum_{l=0}^{L} \alpha(c,l)\delta(t - T_c - \tau_{c,l}), \tag{2.14}$$

where $\alpha(c, l)$ denotes the gain of the lth multipath component in the cth cluster, C is the total number of clusters, and L is the total number of rays within each cluster. The time duration T_c represents the delay of the cth cluster, and $\tau_{c,l}$ is the delay of the lth path in the cth cluster relative to the cluster arrival time. By definition we have $\tau_{c,0} = 0$. The cluster and path arrivals within each cluster are modeled by Poisson processes:

$$p_{T_c}(T_c \mid T_{c-1}) = \lambda \exp[-\Lambda(T_c - T_{c-1})], \qquad c > 0; \tag{2.15}$$

$$p_{\tau_{c,l}}(\tau_{c,l} \mid \tau_{c,l-1}) = \lambda \exp[-\lambda(\tau_{c,l} - \tau_{c,l-1})], \qquad l > 0, \tag{2.16}$$

where Λ is the cluster arrival rate and λ (where $\lambda > \Lambda$) is the ray arrival rate (i.e., the arrival rate of path within each cluster). The path amplitude $|\alpha(c, l)|$ follows the Rayleigh distribution, whereas the phase $\angle\alpha(c, l)$ is distributed uniformly over $[0, 2\pi)$. Specifically, the multipath gain coefficient $\alpha(c, l)$ is modeled as a zero-mean complex Gaussian random variable with variance [Foe03b]

$$\Omega_{c,l} = E[|\alpha(c,l)|^2] = \Omega_{0,0} \exp\left(-\frac{T_c}{\Gamma} - \frac{\tau_{c,l}}{\gamma}\right), \tag{2.17}$$

where $_{0,0}$ is the mean energy of the first path of the first cluster, Γ is the cluster decay factor, and γ is the ray decay factor; this reflects the exponential decay of each cluster as well as the decay of the total cluster power with delay. The four main parameters can be changed for different environments. They provide great flexibility to model very different environments. Figure 2.1 illustrates the various parameters in the S-V model.

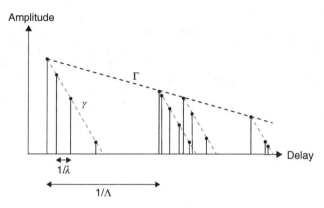

Figure 2.1 Principle of the Saleh–Valenzuela fading model.

2.2.4 Standard UWB Channel Model

In this subsection we describe the UWB channel model that is adopted in the IEEE 802.15.3a standard [TG3a]. The path loss, shadowing, and small-scale fading models of the standard UWB channel are provided below.

- *Path loss.* The path loss specified in the standard is based on free-space path loss, with the center frequency f_c given by $f_c = \sqrt{f_L f_H}$, where f_L and f_H are obtained at the -10-dB edges of the waveform spectrum.
- *Shadowing.* The shadowing is assumed lognormally distributed with standard deviation of 3 dB [i.e., the shadowing is $X_\sigma[\text{dB}] \sim \mathcal{N}(0, \sigma^2)$, with a σ value of 3 dB].
- *Small-scale fading.* The small-scale fading adopted in the IEEE 802.15.3a standard is based on the S-V model. Although the path amplitude $|\alpha(c, l)|$ may follow the lognormal distribution [Foe03b], the Nakagami distribution [Win02], or the Rayleigh distribution [Cra02], the lognormal distribution is adopted in the standard. Four sets of channel model (CM) parameters for different measurement environments were defined: CM1, CM2, CM3, and CM4. CM1 describes a line-of-sight (LOS) scenario with a separation between the transmitter and receiver of less than 4 m. CM2 describes the same range but for a non-LOS situation. CM3

TABLE 2.1 Multipath Channel Model Parameters

Parameters	CM1	CM2	CM3	CM4
Λ (ns^{-1})	0.0233	0.4	0.0667	0.0667
λ (ns^{-1})	2.5	0.5	2.1	2.1
Γ	7.1	5.5	14	24
γ	4.3	6.7	7.9	12

describes a non-LOS scenario for distances of 4 to 10 m between a transmitter and a receiver. CM4 describes an environment with a strong delay dispersion, resulting in a delay spread of 25 ns. Table 2.1 provides the model parameters of CM1 to CM4. Note that the total average received power of the multipath realizations is typically normalized to unity in order to provide a fair comparison with other systems.

3

UWB: SINGLE-BAND APPROACHES

As discussed in Chapter 1, the FCC defines UWB as any signal that occupies at least 500 MHz of bandwidth in the 7.5-GHz spectrum between 3.1 and 10.6 GHz. However, there is no restriction on signal generation techniques to occupy the available UWB spectrum. So far, several UWB transmission techniques have been proposed in the literature. These techniques can be categorized into two major groups: single- and multiband UWB. The single-band approaches, also referred to as *carrier-free* and *impulse communications*, are implemented by direct modulation of information into a sequence of impulselike waveforms which occupy the available bandwidth of 7.5 GHz. Multiple users can be supported via the use of time-hopping or direct-sequence spreading approaches. Multiband approaches, on the other hand, divide the available UWB bandwidth into smaller subbands, each with a bandwidth greater than 500 MHz, to comply with the FCC's definition of UWB signals.

Single-band UWB systems [Sch93, Wel01] have the advantage of low implementation cost. Specifically, since a single-band UWB signal can be transmitted directly without a carrier (carrierless transmission), single-band UWB transceivers do not require mixers or power amplifiers, and the system can be implemented inexpensively. The single-band approaches also have precise position location capability. In particular, the fine time resolution of a single-band UWB signal allows accurate delay estimate, which can be used for precise ranging application and accurate position location. Nevertheless, single-band UWB systems face a challenging problem in building analog circuits and in designing a low-complexity receiver that can capture sufficient multipath energy. Also, as in any UWB system, the performance of single-band UWB is limited by the FCC's constraint on the transmitter power spectral density.

In this chapter we first present modulation and multiple access techniques for single-band UWB systems. Then we describe transmitter signal model and receiver processing under frequency-selective fading channels. To improve system performance under dense multipath environments, the idea of employing multiple-input multiple-output (MIMO) technology in a UWB system has gained considerable

Ultra-Wideband Communications Systems: Multiband OFDM Approach, By W. Pam Siriwongpairat and K. J. Ray Liu
Copyright © 2008 John Wiley & Sons, Inc.

interest recently [Yan02, Kum02, Wei03, Sir04, Sir05a]. By the use of space–time coding [Gue99, Tar98, Ala98], MIMO systems exploit both spatial and temporal diversities, and hence promise to improve system performance significantly. In this chapter we present a general space–time-coded UWB system and provide performance analysis under a Nakagami-m frequency-selective fading environment.

In Section 3.1, we introduce a single-band UWB signal, also known as impulse radio. Modulation techniques, multiple access schemes, and the demodulation techniques are presented in Sections 3.2, 3.3, and 3.4, respectively. In Section 3.5 we provide the signal model and receiver processing for MIMO single-band UWB systems. The performance analysis of single-band UWB approaches is presented in Section 3.6. Simulation results are presented in Section 3.7, and conclusions are drawn in Section 3.8.

3.1 OVERVIEW OF SINGLE-BAND APPROACHES

A traditional UWB technology is based on single-band systems employing carrier-free communications [Sch93, Wel01]. The single-band UWB system is implemented by direct modulation of a sequence of impulselike waveforms (also referred to as *monocycles*) that occupy a bandwidth of several gigahertz. Generally, the pulse duration is on the order of nanoseconds, and the pulse sequence is formed using a single basic pulse shape. The single-band UWB signal is also known as *impulse radio*. Figure 3.1 depicts a sequence of impulselike waveforms with a low duty cycle, where T_f is the frame interval, also known as the pulse repetition time, and T_w is the duration of a monocycle. Typically, T_f is 100 or 1000 times longer than the pulse width T_w.

The monocycle can be any pulse shape whose spectrum satisfies the FCC spectral mask for UWB signals. Pulse shapes that are commonly used in UWB systems include the Gaussian pulse and its higher-order derivatives, the Laplacian pulse, the Rayleigh pulse, and the Hermitian pulse [Fon04]. For instance, a Gaussian pulse is modeled as

$$\tilde{w}(t) = \frac{1}{\sqrt{2\pi}\,\sigma} \exp\left[-\frac{1}{2}\left(\frac{t-\mu}{\sigma}\right)^2 \right], \qquad (3.1)$$

where μ denotes denotes the location of the pulse center and σ is a parameter that determines the width of the pulse. Due to the effect of propagation channel and the variation of antenna characteristics caused by large bandwidth, the shape of

Figure 3.1 Pulse train with a low duty cycle, where T_f is pulse repetition time and T_w is pulse duration.

the monocycle $\tilde{w}(t)$ transmitted is typically modified to its second derivative at the receiver antenna output [Win00, Cra98]. With the pulse transmitted being a Gaussian pulse, the monocycle received can be modeled as the second derivative of the Gaussian pulse [Win00]:

$$w(t) = \frac{1}{\sqrt{2\pi}\,\sigma} \left[1 - \left(\frac{t-\mu}{\sigma} \right)^2 \right] \exp\left[-\frac{1}{2}\left(\frac{t-\mu}{\sigma} \right)^2 \right], \tag{3.2}$$

whose pulse duration is $T_w = 7\sigma$ and pulse center is at $\mu = 3.5\sigma$.

At the transmitter the information bits are modulated into the sequence of the monocycles. At the receiver, the resolvable multipath components are combined such that the information is recovered with less probability of error. A detailed description of an impulse-radio transceiver model is provided below.

3.2 MODULATION TECHNIQUES

In single-band UWB systems, the information can be modulated into the amplitudes, phases, or positions of the pulses. The modulation schemes for single-band UWB approaches include pulse amplitude modulation (PAM), on–off keying (OOK), phase shift keying (PSK), and pulse position modulation (PPM) [Sch93, Wel01].

3.2.1 Pulse Amplitude Modulation

The information in a PAM signal is conveyed in the amplitude of pulses. Specifically, an M-ary PAM signal comprises a sequence of modulated pulses with M different amplitude levels. The PAM signal can be modeled as

$$\tilde{x}(t) = \sum_{k=-\infty}^{\infty} a_m(k)\tilde{w}(t - kT_f), \tag{3.3}$$

where $a_m(k)$ is the amplitude of the kth pulse, which depends on the M-ary information symbol $m \in \{0, 1, \ldots, M - 1\}$. Figure 3.2 shows a 4-ary PAM signal. A PAM signal is simple to generate; however, it is vulnerable to channel noise, which can change the pulse amplitude and cause false detection. Moreover, since the pulse transmitted is periodic, it produces discrete lines on the power spectral density of a PAM signal. Such discrete spectral lines can cause interference to systems sharing a frequency spectrum.

Figure 3.2 Four-ary PAM signal.

Figure 3.3 OOK signal.

3.2.2 On–Off Keying

OOK is a special case of PAM with binary symbol $m \in \{0, 1\}$ and pulse amplitude $a_m(k) = m(k)$. In other words, a pulse is transmitted if the information bit is 1, while it is absent if the information bit is 0. An OOK signal can be modeled as

$$\tilde{x}(t) = \sum_{k=-\infty}^{\infty} m(k)\tilde{w}(t - kT_f). \tag{3.4}$$

Figure 3.3 depicts an OOK signal. An OOK system is the simplest system to implement, but it yields poor performance since noise and interference can easily cause false detection.

3.2.3 Phase Shift Keying

In binary PSK (BPSK) or biphase modulation, the binary data are carried in the polarity of the pulses. For example, a pulse has positive polarity if the information bit is 1, whereas it has negative polarity if the information bit is 0. A BPSK signal can be modeled as

$$\tilde{x}(t) = \sum_{k=-\infty}^{\infty} d(k)\tilde{w}(t - kT_f), \tag{3.5}$$

where

$$d(k) = \begin{cases} 1 & \text{if information bit is 1;} \\ -1 & \text{if information bit is 0} \end{cases} \tag{3.6}$$

is the polarity of the modulated pulse. Figure 3.4 depicts a BPSK signal. The BPSK signal yields better performance than the OOK signal since the different pulse level is twice the pulse amplitude. The BPSK signal also has fewer discrete lines on the power spectral density than does the PAM signal since the change in the polarity of the pulses results in a zero mean.

Figure 3.4 Binary PSK signal.

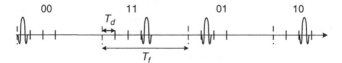

Figure 3.5 Four-ary PPM signal.

3.2.4 Pulse Position Modulation

PPM is one of the most popular modulation techniques in UWB literature. With PPM, the information is carried in the fine time shift of the pulse. The M-ary PPM signal can be modeled as

$$\tilde{x}_u(t) = \sum_{k=-\infty}^{\infty} \tilde{w}(t - kT_f - m(k)T_d), \tag{3.7}$$

where $m(k) \in \{0, 1, \ldots, M - 1\}$ is the kth M-ary symbol and T_d is the modulation delay, which provides a pulse time shift to represent each M-ary symbol. Specifically, the monocycle conveying information $m(k)$ is shifted by a modulation delay of $m(k)T_d$ seconds. A 4-ary PPM signal is illustrated in Fig. 3.5.

Since the PPM signal carries information in the time shift of the pulses, it is less sensitive to noise than are PAM or PSK signals, which carry information in the pulse amplitude. Moreover, the pseudorandom code sequence of the pulse positions reduces the discrete lines on the power spectral density of the PPM signal more than those of the PAM signal.

3.3 MULTIPLE ACCESS TECHNIQUES

In single-band UWB systems, multiple users share a single UWB spectrum simultaneously. To accommodate these multiple users, proper multiple access techniques are necessary. Two commonly used multiple techniques in single-band UWB systems are time-hopping (TH) and direct-sequence (DS) spreading techniques. In TH-based systems [Sch93, Win98, Wel01], the information is sent with a time offset for each pulse determined by the TH sequence. In DS spreading systems [Foe02a, Bou03], the data are carried in multiple pulses whose amplitudes are based on a certain spreading code. TH and DS spreading codes both provide robustness against multiuser interference. The performance comparisons of TH and DS schemes for single-antenna systems have been studied [Som02, Dur03a] and it has been shown that TH-UWB systems are suitable in theory and analysis but are seldom, if ever, used in practice (e.g., see the IEEE 802.15.3a standards process [TG3a]). On the other hand, DS-UWB has been shown to be a promising scheme for single-carrier UWB communications. The TH-UWB and DS-UWB systems are described in the following subsections.

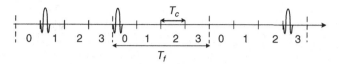

Figure 3.6 Pulse train with TH sequence $\{1, 0, 3, \ldots\}$.

3.3.1 Time-Hopping UWB

TH-UWB utilizes low-duty-cycle pulses, where the time spreading between the pulses is used to provide time multiplexing of users. Basically, each frame interval of duration is divided into multiple smaller segments; only one of these segments carries the user's transmitted monocycle. A unique code, also referred to as a *TH sequence*, is assigned to each user to specify which segment in each frame interval is used for transmission. As shown in Fig. 3.6, the frame interval T_f is divided into N_c segments of T_c seconds where $N_c T_c \leq T_f$. The TH sequence is denoted by $\{c(k)\}$, $0 \leq c(k) \leq N_c - 1$. It provides an additional time shift of $c(k)T_c$ seconds to the kth monocycle to allow multiple access without catastrophic collisions.

Incorporating the TH sequence to the pulse train, we have

$$\tilde{x}(t) = \sum_{k=-\infty}^{\infty} \tilde{w}(t - kT_f - c(k)T_c). \tag{3.8}$$

Figure 3.6 is a pulse train with TH sequence $c(k) = \{1, 0, 3, \ldots\}$. In a synchronized network, an orthogonal TH sequence that satisfies $c^{(u)}(k) \neq c^{(u')}(k)$ for all k's and for any two users $u \neq u'$ can be adopted to minimize interference between the users. The performance of synchronous multiple access systems using various TH sequences such as the Gold sequence and a simulated annealing code has been studied [Can03]. For an asynchronous system, the choice of orthogonal TH sequence does not guarantee collision-free transmission [Dur03a].

The TH technique can be used with PAM, PSK, or PPM modulations. As in Section 3.2, let $m(k) \in \{0, 1, \ldots, M - 1\}$ denote the M-ary symbol. Then, the TH-UWB signal with PAM modulation can be modeled as

$$\tilde{x}(t) = \sum_{k=-\infty}^{\infty} a_m(k)\tilde{w}(t - kT_f - c(k)T_c), \tag{3.9}$$

where $a_m(k)$ is the amplitude of the monocycle carrying information $m(k)$. Similarly, a TH-UWB signal with BPSK modulation is given by

$$\tilde{x}(t) = \sum_{k=-\infty}^{\infty} d(k)\tilde{w}(t - kT_f - c(k)T_c), \tag{3.10}$$

where $d(k)$ is the polarity of the modulated pulse as specified in (3.6). With PPM modulation, the information is conveyed by the position of the pulses. A TH-UWB

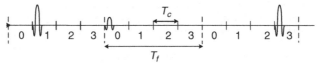

TH-UWB signal with PAM modulation

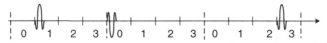

TH-UWB signal with PSK modulation

TH-UWB signal with PPM modulation

Figure 3.7 TH-UWB signals with various modulation.

signal with M-ary PPM modulation can then be described as

$$\tilde{x}(t) = \sum_{k=-\infty}^{\infty} \tilde{w}(t - kT_f - c(k)T_c - m(k)T_d), \tag{3.11}$$

where T_d is the modulation delay. Figure 3.7 shows TH-UWB signals with BPAM, BPSK, and BPPM modulations.

3.3.2 Direct-Sequence UWB

DS-UWB employs a train of high-duty-cycle pulses whose polarities follow pseudo-random code sequences. Specifically, each user in the system is assigned a pseudo-random sequence that controls pseudorandom inversions of the UWB pulse train.

In a DS-UWB system with BPSK modulation, the binary symbol $d(k)$ to be transmitted over the kth frame interval is spread by a sequence of multiple monocycles $\{c(n_c)\tilde{w}(t - kT_f - n_cT_c)\}_{n_c=0}^{N_c-1}$, whose polarities are determined by the spreading sequence $\{c(n_c)\}_{n_c=0}^{N_c-1}$. Such a spreading sequence is assigned uniquely to each user in a multiple access system in order to allow multiple transmissions with little interference. Similar to the TH system, an orthogonal spreading sequence such as a Gold sequence or Hadamard–Walsh code can be selected to mitigate multiple access interference in a synchronous network [Foe02a].

The DS-BPSK signal transmitted can be described as [Foe02a, Bou03]

$$\tilde{x}(t) = \frac{1}{\sqrt{N_c}} \sum_{k=-\infty}^{\infty} d(k) \sum_{n_c=0}^{N_c-1} c(n_c)\tilde{w}(t - kT_f - n_cT_c), \tag{3.12}$$

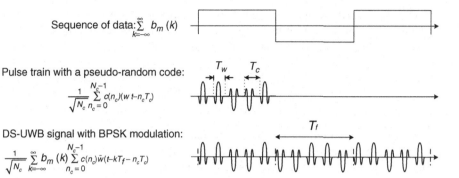

Figure 3.8 DS-UWB signal with BPSK modulation.

where $d(k) \in \{-1, 1\}$ is the modulated binary data [i.e., $d(k) = -1$ when the data bit is 0 and $d(k) = 1$ when the data bit is 1]. In (3.12), the sequence $\{c(n_c)\}_{n_c=0}^{N_c-1} \in \{-1, 1\}$ represents the pseudorandom code or spreading sequence and $T_c \geq T_w$ denotes the hop period. The factor $1/\sqrt{N_c}$ is introduced such that the sequence of N_c monocycles has unit energy. An example of a DS-UWB signal is shown in Fig. 3.8.

3.4 DEMODULATION TECHNIQUES

In this section we first present the received signal model and then discuss two detection techniques in single-band UWB systems: the correlation and RAKE receivers.

3.4.1 Received Signal Model

Consider a multiuser UWB system with N_u users. The uth user transmits a single-band UWB signal $\tilde{x}^{(u)}(t)$ ($u = 1, 2, \ldots, N_u$) carrying information sequence $d^{(u)}(k) \in \{-1, 1\}$, as described in Section 3.3. Let the channel impulse response of the uth user be modeled by a frequency-selective channel model:

$$h^{(u)}(t) = \sum_{l=0}^{L^{(u)}-1} \alpha^{(u)}(l)\delta\big(t - \tau^{(u)}(l)\big), \tag{3.13}$$

where the superscript u indicates the user u, $\alpha^{(u)}(l)$ represents the multipath gain coefficients, $L^{(u)}$ denotes the number of resolvable paths, and $\tau^{(u)}(l)$ represents the path delays relative to the delay of the desired user's first arrival path.

The signal received consists of multipath signals from all active users and thermal noise. Due to the effect of propagation channel and the variation of antenna characteristics caused by large bandwidth, the shape of the monocycle received is different from that of the monocycle transmitted. Typically, the monocycle transmitted is modified to its second derivative at the receiver antenna output [Win00]. For notational convenience, we denote the received monocycle waveform by $w(t)$ and let $x^{(u)}(t)$ be of the form similar to the transmitted waveform $\tilde{x}^{(u)}(t)$ by replacing $\tilde{w}(t)$ with $w(t)$. Accordingly, the signal received can be modeled as

$$r(t) = \sum_{u=1}^{N_u} \sum_{l=0}^{L^{(u)}-1} \alpha^{(u)}(l) x^{(u)}(t - \tau^{(u)}(l)) + n(t), \qquad (3.14)$$

where $n(t)$ is the additive noise, which is modeled as a real additive white Gaussian noise process with zero mean and two-sided power spectral density $N_0/2$.

In the following subsections we describe the techniques used to detect the information in the received signal $r(t)$. Without loss of generality, we let user 1 be the desired user and assume that $\tau^{(1)}(0) = 0$.

3.4.2 Correlation Receiver

The correlation receiver is the optimum receiver for a single bit of a binary-modulated single-band UWB signal in an additive white Gaussian noise (AWGN) channel. The signal received at the correlation receiver is correlated with the pulse expected, called a reference or template signal, and binary decisions are made depending on the sign of correlation values. Figure 3.9 depicts a correlator with a reference signal $v_{k'}^{(1)}(t)$. The reference signal depends on modulation and multiple access techniques as follows.

- *TH-BPPM.* The design of a reference signal for a TH-BPPM system depends on the choice of the modulation delay, T_d. Any choice of $T_d \geq T_w$ yields a design equivalent to that of an orthogonal signaling scheme. However, due to the multipath propagation, such orthogonality can be corrupted at the receiver. To preserve the orthogonality, T_d has to satisfy the condition $T_d \geq \max_u\{\tau^{(u)}(L^{(u)} - 1)\} + T_w$. This results in a loss of the transmission rate. In the following we choose T_d to minimize the correlation $\int_{-\infty}^{\infty} w(t)w(t - T_d)dt$, as in [Sch93]. With this choice of T_d, the design is close to an antipodal signaling scheme, and the transmission rate can be made equal to that of a system with BPSK modulation. In this case the reference signal is given by

$$v_{k'}^{(1)}(t) = w(t - k'T_f - c^{(1)}(k')T_c) - w(t - k'T_f - c^{(1)}(k')T_c - T_d). \quad (3.15)$$

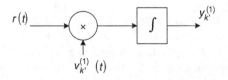

Figure 3.9 Correlator with reference signal $v_{k'}^{(1)}(t)$.

- *TH-BPSK*. The reference waveform used in a TH-BPSK system is the delayed version of the monocycle received:

$$v_{k'}^{(1)}(t) = w\big(t - k'T_f - c^{(1)}(k')T_c\big). \tag{3.16}$$

- *DS-BPSK*. The DS-BPSK reference signal is a sequence of pulses whose polarities are modulated with the design user's spreading code. When user 1 is the design user, the spreading sequence is $\{c^{(1)}(n_c)\}$, and the corresponding reference signal is given by

$$v_{k'}^{(1)}(t) = \frac{1}{\sqrt{N_c}} \sum_{n_c'=0}^{N_c-1} c^{(1)}(n_c')w(t - k'T_f - n_c'T_c). \tag{3.17}$$

The output of the correlator is given by

$$y_{k'}^{(1)} = \int_{-\infty}^{\infty} v_{k'}^{(1)}(t)r(t)\,dt. \tag{3.18}$$

Consider an AWGN channel in the absence of multiuser interference. The signal received comprises simply the signal transmitted and the additive noise [i.e., $r(t) = x^{(1)}(t) + n(t)$]. In this case, the output of the correlator can be simplified to

$$y_{k'}^{(1)} = \int_{-\infty}^{\infty} v_{k'}^{(1)}(t)x^{(1)}(t)\,dt + \int_{-\infty}^{\infty} v_{k'}^{(1)}(t)n(t)\,dt = (1 - \rho)d^{(1)}(k') + n_{k'}, \tag{3.19}$$

where $\rho = 0$ for BPSK modulation and $\rho = \int_{-\infty}^{\infty} w(t - T_d)w(t)ds$ for BPPM modulation. In (3.19), $n_{k'}$ is the sample noise, which is a zero-mean Gaussian random variable with variance $N_0/2$. From (3.19), the maximum likelihood decision rule is given by

$$\hat{d}^{(1)}(k') = \begin{cases} 1 & \text{if } y_{k'}^{(1)} \geq 0; \\ -1 & \text{otherwise.} \end{cases} \tag{3.20}$$

Clearly, the correlation receiver is optimal in an AWGN channel. However, the UWB channel is typically a frequency-selective fading channel and the correlation receiver does not perform well in this case.

3.4.3 RAKE Receiver

In a frequency-selective fading channel, a RAKE receiver can be used to exploit multipath diversity by combining constructively the monocycles received from resolvable multipath components. A typical RAKE receiver is composed of several correlators followed by a linear combiner, as shown in Fig. 3.10. The signal received at the RAKE receiver is correlated with delayed versions of the reference pulse $\{v_{k'}^{(1)}(t - l)\}_{l=0}^{L-1}\}$, multiplied by the tap weights $\{g^{(1)}(l)\}_{l=0}^{L-1}$, and finally, combined linearly. The reference pulse is the same as that defined in Section 3.4.2. The performance of a RAKE receiver depends on the path selection technique and the combining method. Various

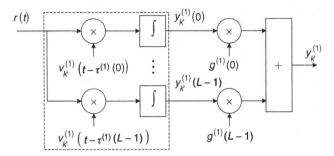

Figure 3.10 RAKE receiver with reference signals $v_{k'}^{(1)}(t - \tau^{(1)}(l))$.

combinations of path selection and combining methods have been studied for the trade-off between the receiver complexity and system performance.

Path Selection Techniques. In general, the number of resolvable path components in a realistic dense multipath channel (i.e., an indoor channel) is approximately proportional to the transmission bandwidth BW and the excess delay T_D of the channel. Since the product BW $\cdot T_D$ is likely to be large for a UWB system, an all-RAKE receiver that combines all resolvable paths, although it provides high performance, may not be realizable and is expensive to implement. Complexity and performance issues have motivated studies of reduced-complexity RAKE receivers that process only a subset of the resolved multipath components available. Three of the path selection techniques proposed in the UWB literature are described briefly below [Cas02]. Let the number of resolvable path components be L_1 and the number of paths chosen to be combined be L, $L \leq L_1$.

- *Maximum selection.* Select the L best paths of L_1 resolvable paths (i.e., the receiver makes the best use of its L available taps). To select properly requires keeping track of all L_1 path components, using algorithms to sort all these L paths by the magnitude of their instantaneous path gains, which would require instantaneous and highly accurate channel estimation.
- *Partial selection.* Select the first nonzero L arriving paths, which are not necessarily the best. The partial selection requires neither path amplitude knowledge nor the selection mechanism. This allows a partial selection receiver to have less complexity than a maximum selection receiver but at the cost of lower performance. It has been shown [Cas02] that the performance loss of a partial selection technique compared to maximum selection is quite small in a Nakagami fading channel but larger in a Rayleigh fading channel.
- *Threshold selection.* Select the first L paths in which the magnitude of the path gains is greater than a threshold. Like partial selection, this technique does not require either a sorting algorithm or amplitude knowledge; however,

the proper threshold needs to be defined. Different thresholds may result in different performance.

Combining Methods. The combining method specifies the choice of tap weights to be used.

- *Equal gain combining* (EGC) [Pro01]. The outputs of the correlators are summed together directly and fed to the detector [i.e., $g^{(1)}(l) = 1$, for all l]. It is the simplest form of combiner that does not require any knowledge of the path amplitudes.

- *Maximum ratio combining* (MRC) [Pro01]. The outputs of the correlators are weighted in direct proportion to the signal strength received (the square root of the power level) and then fed to the detector. This requires an estimation of the amplitude parameters of each path chosen. The RAKE receiver using MRC maximizes the system's instantaneous signal-to-noise ratio (SNR) when no narrowband interference exists. Its performance degrades in the presence of narrowband interference [Ber02].

- *Minimum mean-square-error combining* (MMSEC) [Ber02]. The objective of MMSEC is to achieve narrowband interference suppression. The tap weights for this technique are chosen to maximize instantaneous SNR in a the presence of narrowband interference. An adaptive MMSE algorithm can also be used to improve interference suppression in a time-varying environment. Compared to EGC and MRC, MMSEC has the highest complexity, but it is likely to have the best performance when narrowband interference exists.

3.5 MIMO SINGLE-BAND UWB

MIMO space–time-coded systems are well known for their effectiveness at improving system performance under multipath scenarios. A key concept to approaching such improvement is space–time coding [Gue99, Tar98, Ala99, Tar99, Hoc00], which is based on introducing joint processing in time as well as in space via the use of multiple spatially distributed antennas. Through this approach, MIMO can provide diversity and coding gains simultaneously and hence yield high spectral efficiency and remarkable quality improvement.

To exploit the advantages of the MIMO technique in single-band UWB systems, space–time-coded UWB systems have been studied [Yan02, Kum02, Wei03, Sir04, Sir05a]. In this section we describe MIMO space–time-coded systems and then present space–time-coded UWB systems with various modulation and multiple access schemes, including TH-BPPM, TH-BPSK, and DS-BPSK.

3.5.1 MIMO Space–Time-Coded Systems

In a point-to-point MIMO system, multiple antennas are deployed at both transmitter and receiver, as shown in Fig. 3.11. At the transmitter, the data sequence is divided into

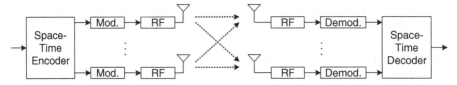

Figure 3.11 MIMO space–time-coded system.

blocks. Each block is encoded into a codeword matrix. Each column of the codeword matrix contains a sequence of symbols that will be sent from each transmitting antenna over a series of time-slot or frequency tones (depending on the modulation scheme, e.g., single carrier or multicarrier). These symbol streams are modulated with a pulse-shaping function, translated to the passband via parallel radio-frequency (RF) chains, and then transmitted simultaneously over all transmitting antennas. After down-conversion, matched-filtering, and demodulation processes, the receiver decodes signals jointly across all receiver antennas.

The codeword matrix was originally designed for flat fading channels. In this case, the row and column indices of the codeword matrix indicate the dimensions of space and time and hence the codeword matrix is called the *space–time-code*. Consider a MIMO system employing N_t transmitter and N_r receiver antennas. Let K denote the number of time slots for one codeword transmission; then information is encoded jointly across N_t transmitter antennas and K time slots. The corresponding space–time codeword matrix is an $K \times N_t$ matrix **D** whose (k, i)th element, $d_i(k)$, represents the symbol transmitted at the transmitter antenna i over time slot k. The MIMO channel is described by an $N_t \times N_r$ matrix **A**. The (i, j)th component of **A**, denoted by α_{ij}, is the channel fading coefficient from the ith transmitter to the jth receiver antenna. The signal received at each receiver antenna is a noisy superposition of the N_t transmitted signals degraded by the channel fading. Consider the case when the channels are quasistatic (i.e., they remain constant during the transmission of an entire codeword). The signal received can be described as an $K \times N_r$ matrix:

$$\mathbf{Y} = \sqrt{\frac{E}{N_t}} \mathbf{D} \mathbf{A} + \mathbf{N}, \qquad (3.21)$$

where **N** is the matrix of additive complex Gaussian noises, each with zero mean and variance $N_0/2$ per dimension. For normalization purposes, the fading coefficient for each transmitter–receiver link is assumed to have unit variance, and the space–time code satisfies the energy constraint $E[\|\mathbf{D}\|^2] = KN_t$. Here $E[\mathbf{X}]$ and $\|\mathbf{X}\|$ denote the expectation and Frobenius norm[1] of **X**, respectively. The factor $\sqrt{1/N_t}$

---

[1] The Frobenius norm of an $M \times N$ matrix $\mathbf{X} = (x_{mn})$ is defined as [Hor85].

$$\|\mathbf{X}\|^2 = \mathrm{tr}(\mathbf{X}^{\mathcal{H}} \mathbf{X}) = \mathrm{tr}(\mathbf{X} \mathbf{X}^{\mathcal{H}}) = \left(\sum_{m=0}^{M-1} \sum_{n=0}^{N-1} |x_{mn}|^2 \right). \qquad (3.22)$$

ensures transmitter energy identical to single-antenna transmission. Assuming that the channel-state information (CSI) is known perfectly at the receiver, the receiver performs maximum likelihood decoding by choosing the codeword $\hat{\mathbf{D}}$ that minimizes the square Euclidean distance between the hypothesized and actual received signal matrices, that is,

$$\hat{\mathbf{D}} = \underset{\mathbf{D}}{\arg\min} \left\| \mathbf{Y} - \sqrt{\frac{E}{N_t}} \mathbf{DA} \right\|^2 . \tag{3.23}$$

The upper bound of the average pairwise error probability (PEP) between $\mathbf{D}$ and $\hat{\mathbf{D}}$ is of the form [Gue99, Tar98]

$$P(\mathbf{D} \rightarrow \hat{\mathbf{D}}) \leq \left[G_c \frac{\rho}{4N_t} \right]^{-G_d} , \tag{3.24}$$

where $\rho = E/N_0$ is the average SNR at each receiver antenna. The quantities G_d and G_c depend on the distribution of channel fading coefficients and the structure of $\mathbf{D}$. They characterize the performance of space–time-coded MIMO system as follows. The exponent G_d determines the slope of the error probability curve plotted as a function of SNR (measured in decibels). The factor G_c displaces the performance curve rather than alternating its slope. The minimum values of G_d and G_c over all pairs of distinct codewords are called *diversity gain* and *coding gain*, respectively.

3.5.2 Space–Time-Coded UWB Systems

In this section we describe the transmitter signal model and receiver processing for space–time-coded UWB systems with TH-BPPM, TH-BPSK, and DS-BPSK [Sir04, Sir05a]. We consider a space–time-coded UWB system with N_u users, each equipped with N_t transmitter antennas and a receiver with N_r receiver antennas, as shown in Fig. 3.12. For each user, the sequence of binary symbols is divided into blocks of N_b

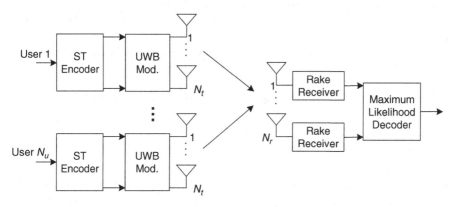

Figure 3.12 Multiuser space–time-coded UWB system.

symbols. Each block is encoded into a space–time codeword to be sent simultaneously over all transmitter antennas. The space–time codeword matrix can be expressed as an $K \times N_t$ matrix

$$
\mathbf{D}^{(u)} = \begin{pmatrix} d_1^{(u)}(0) & d_2^{(u)}(0) & \cdots & d_{N_t}^{(u)}(0) \\ d_1^{(u)}(1) & d_2^{(u)}(1) & \cdots & d_{N_t}^{(u)}(1) \\ \vdots & \vdots & \ddots & \vdots \\ d_1^{(u)}(K-1) & d_2^{(u)}(K-1) & \cdots & d_{N_t}^{(u)}(K-1) \end{pmatrix}, \tag{3.25}
$$

where $d_i^{(u)}(k) \in \{-1, 1\}$ represents the binary symbol transmitted by the uth user at transmitter antenna i over time slot k. Since K time slots are required for N_b symbol transmission, the code rate is $R = N_b/K$. The transmitter converts the space–time codeword matrix into an $K \times N_t$ UWB space–time signal matrix $\tilde{\mathbf{X}}^{(u)}(t) = [\tilde{x}_i^{(u)}(k;t)]$ by modulating the symbol $d_i^{(u)}(k)$ into a UWB signal $\tilde{x}_i^{(u)}(k;t)$. The UWB signal depends on the particular multiple access and modulation techniques. With the TH approach, $\tilde{x}_i^{(u)}(k;t)$ is modeled as [see (3.7)]

$$
\tilde{x}_i^{(u)}(k;t) = \sqrt{\frac{E_u}{N_t}} \tilde{w} \left(t - kT_f - c^{(u)}(k)T_c - \frac{1 - d_i^{(u)}(k)}{2}T_d \right) \tag{3.26}
$$

for BPPM and

$$
\tilde{x}_i^{(u)}(k;t) = \sqrt{\frac{E_u}{N_t}} d_i^{(u)}(k) \tilde{w} \left(t - kT_f - c^{(u)}(k)T_c \right) \tag{3.27}
$$

for BPSK. With the DS-BPSK technique [see (3.12)], the UWB signal becomes

$$
\tilde{x}_i^{(u)}(k;t) = \sqrt{\frac{E_u}{N_t N_c}} d_i^{(u)}(k) \sum_{n_c=0}^{N_c-1} c^{(u)}(n_c) \tilde{w}(t - kT_f - n_c T_c). \tag{3.28}
$$

Note that the factor $\sqrt{1/N_t}$ is introduced to ensure that the total energy transmitted by the uth user is E_u during each frame interval, independent of the number of transmitter antennas.

Let the channel impulse response from transmitter antenna i of user u to receive antenna j be modeled as

$$
h_{ij}^{(u)}(t) = \sum_{l=0}^{L^{(u)}-1} \alpha_{ij}^{(u)}(l) \delta \left(t - \tau^{(u)}(l) \right), \tag{3.29}
$$

where $\alpha_{ij}^{(u)}(l)$ is the multipath gain coefficients, $L^{(u)}$ denotes the number of resolvable paths, and $\tau^{(u)}(l)$ represents the path delays relative to the delay of the desired user's first arrival path. For simplicity, we assume that the minimum resolvable delay is equal to the pulse width, as in [Foe02a]. This assumption implies that $\tau^{(u)}(l) - \tau^{(u)}(l-1) \geq T_w$ for any $l \in \{2, 3, \ldots, L^{(u)}\}$.

With the first user being the desired user, the signal received at receiver antenna j can be expressed as

$$r_j(t) = y_j^{(1)}(t) + n_j^{mui}(t) + n_j(t), \tag{3.30}$$

where $n_j(t)$ is the additive white Gaussian noise process, $n_j^{mui}(t) = \sum_{u=2}^{N_u} y_j^{(u)}(t)$ is the multiuser interference, and

$$y_j^{(u)}(t) = \sum_{i=1}^{N_t} \sum_{k=0}^{K-1} \sum_{l=0}^{L^{(u)}-1} \alpha_{ij}^{(u)}(l) x_i^{(u)}\big(k; t - \tau^{(u)}(l)\big) \tag{3.31}$$

represents the signal from the uth user. The real additive white Gaussian noise process, $n_j(t)$, is assumed to have zero mean and two-sided power spectral density $N_0/2$. By choosing the signal parameters such that

$$N_c T_c + \max_u \big\{\tau^{(u)}\big(L^{(u)}\big)\big\} \le T_f, \tag{3.32}$$

we can guarantee that there is no intersymbol interference (ISI).

As depicted in Fig. 3.12, a single-band UWB MIMO receiver comprises N_r RAKE receivers and a maximum likelihood decoder where the decoding process is performed jointly across all N_r receiver antennas. The RAKE receiver employs L ($L \le \max_u\{L^{(u)}\}$) correlators, each using delayed versions of the received monocycle as the reference waveform. The reference waveform $v_{k'}^{(1)}(t)$ is in the same form as that presented in Section 3.4.2. The output of the l'th correlator at receiver antenna j is given by

$$y_{j,k'}^{(1)}(l') = \int_{-\infty}^{\infty} v_{k'}^{(1)}\big(t - \tau^{(1)}(l')\big) r_j(t) \, dt. \tag{3.33}$$

Substituting the receiver signal in (3.30) into (3.33), we have

$$y_{j,k'}^{(1)}(l') = \sum_{i=1}^{N_t} \sum_{k=0}^{K-1} \sum_{l=0}^{L^{(1)}-1} \alpha_{ij}(l) \int_{-\infty}^{\infty} v_{k'}^{(1)}\big(t - \tau^{(1)}(l')\big) x_i^{(1)}\big(k; t - \tau^{(u)}(l)\big) \, dt$$

$$+ \int_{-\infty}^{\infty} v_{k'}^{(1)}\big(t - \tau^{(1)}(l')\big) n_j^{mui}(t) \, dt + \int_{-\infty}^{\infty} v_{k'}^{(1)}\big(t - \tau^{(1)}(l')\big) n_j(t) \, dt$$

$$\stackrel{\Delta}{=} y_{j,k'}^d(l') + \underbrace{n_{j,k'}^{mui}(l') + n_{j,k'}(l')}_{n_{j,k'}^{tot}(l')}, \tag{3.34}$$

where $v_{j,k'}^d(l')$, $n_{j,k'}^{mui}(l')$, and $n_{j,k'}(l')$ denote the correlator outputs corresponding to the desired transmitted information, multiuser interference, and thermal noise, respectively. The term $n_{j,k'}^{tot}(l') \stackrel{\Delta}{=} n_{j,k'}^{mui}(l') + n_{j,k'}(l')$ is the combination of multiuser interference and sample noise. Assuming no ISI [i.e., (3.32) is satisfied], only the desired user's signal transmitted during the k'th frame will contribute to $y_{j,k'}^d(l')$. Thus, we can express $y_{j,k'}^{d}(l')$ in (3.34) as

$$y_{j,k'}^{d}(l') = \sum_{i=1}^{N_t} \sum_{l=0}^{L^{(1)}-1} \alpha_{ij}^{(1)}(l) \int_{-\infty}^{\infty} v_{k'}^{(1)}\left(t - \tau^{(1)}(l')\right) x_i^{(1)}\left(k'; t - \tau^{(1)}(l)\right) dt. \quad (3.35)$$

The correlator outputs depends on multiple access and modulation techniques as follows.

For notational convenience, we denote the autocorrelation function of $w(t)$ as

$$\gamma(s) = \int_{-\infty}^{\infty} w(t - s)w(t)\,ds, \quad (3.36)$$

where $\gamma(0) = 1$ and $\gamma(s)$ can be approximately zero when $|s| \geq T_w$ (i.e., the time difference between the monocycles is longer than the pulse duration).

- *TH-BPPM*. The correlator output can be derived by substituting the transmitted TH-BPPM signal in (3.26) and the reference signal in (3.15) into (3.33). After some manipulation, we obtain the correlator output of TH-BPPM system as [Sir05a]

$$y_{j,k'}^{(1)}(l') = [1 - \gamma(T_d)]\sqrt{\frac{E_1}{N_t}} \sum_{i=1}^{N_t} d_i^{(1)}(k)\alpha_{ij}^{(1)}(l') + n_{j,k'}^{\text{tot}}(l'). \quad (3.37)$$

Next, we arrange the correlator outputs in matrix form as

$$\mathbf{Y}_j = [1 - \gamma(T_d)]\sqrt{\frac{E_1}{N_t}} \mathbf{D}^{(1)}\mathbf{A}_j^{(1)} + \mathbf{N}_j^{\text{tot}}, \quad (3.38)$$

where $\mathbf{D}^{(1)}$ is the desired user's space–time symbol defined in (3.25). Matrices $\mathbf{Y}_j$ and $\mathbf{N}_j^{\text{tot}}$ are both of size $K \times L$, whose (k, l)th elements are $y_{j,k'}^{(1)}(l)$ and $n_{j,k'}^{\text{tot}}(l)$, respectively. The multipath gain coefficient matrix $\mathbf{A}_j^{(1)}$ of size $N_t \times L$ is formatted as

$$\mathbf{A}_j^{(1)} = \begin{pmatrix} \alpha_{1j}^{(1)}(0) & \alpha_{1j}^{(1)}(1) & \cdots & \alpha_{1j}^{(1)}(L-1)) \\ \alpha_{2j}^{(1)}(0) & \alpha_{2j}^{(1)}(1) & \cdots & \alpha_{2j}^{(1)}(L-1)) \\ \vdots & \vdots & \ddots & \vdots \\ \alpha_{N_tj}^{(1)}(0) & \alpha_{N_tj}^{(1)}(1) & \cdots & \alpha_{N_tj}^{(1)}(L-1)) \end{pmatrix}. \quad (3.39)$$

Given the CSI on MIMO channels, the decoder performs maximum likelihood decoding by selecting a codeword $\hat{\mathbf{D}}^{(1)}$ that minimizes the square Euclidean distance between the hypothesized and actual correlator output matrices. The decision rule can be stated as

$$\hat{\mathbf{D}}^{(1)} = \underset{\mathbf{D}^{(1)}}{\operatorname{argmin}} \sum_{j=1}^{N_r} \left\| \mathbf{Y}_j - [1 - \gamma(T_d)] \sqrt{\frac{E_1}{N_t}} \mathbf{D}^{(1)} \mathbf{A}_j^{(1)} \right\|^2 . \tag{3.40}$$

- *TH-BPSK*. Substituting (3.27) and (3.16) into (3.33), we obtain

$$y_{j,k'}^{(1)}(l') = \sqrt{\frac{E_1}{N_t}} \sum_{i=1}^{N_t} d_i^{(1)}(k') \alpha_{ij}^{(1)}(l') + n_{j,k'}^{\text{tot}}(l'). \tag{3.41}$$

The correlator outputs can be written in matrix form as

$$\mathbf{Y}_j = \sqrt{\frac{E_1}{N_t}} \mathbf{D}^{(1)} \mathbf{A}_j^{(1)} + \mathbf{N}_j^{\text{tot}}, \tag{3.42}$$

in which $\mathbf{Y}_j$, $\mathbf{A}_j^{(1)}$, and $\mathbf{N}_j^{\text{tot}}$ are in the same form as those stated in (3.38). The decision rule for the maximum likelihood decoder can be written similar to (3.40) as

$$\hat{\mathbf{D}}^{(1)} = \underset{\mathbf{D}^{(1)}}{\operatorname{argmin}} \sum_{j=1}^{N_r} \left\| \mathbf{Y}_j - \sqrt{\frac{E_1}{N_t}} \mathbf{D}^{(1)} \mathbf{A}_j^{(1)} \right\|^2 . \tag{3.43}$$

- *DS-BPSK*. The DS-BPSK receiver adopts the monocycle sequence in (3.17) as the reference waveform. From (3.28), (3.17), and (3.33), we have

$$y_{j,k'}^{(1)}(l') = \sqrt{\frac{E_1}{N_t}} \sum_{i=1}^{N_t} d_i^{(1)}(k') \sum_{l=0}^{L^{(1)}-1} \alpha_{ij}^{(1)}(l) f(l, l') + n_{j,k'}^{\text{tot}}(l'), \tag{3.44}$$

where

$$f(l, l') \triangleq \frac{1}{N_c} \sum_{n_c'=0}^{N_c-1} c^{(1)}(n_c') \sum_{n_c=0}^{N_c-1} c^{(1)}(n_c) \gamma \left[(n_c - n_c') T_c + \tau^{(1)}(l) - \tau^{(1)}(l') \right]. \tag{3.45}$$

We rewrite the correlator outputs in matrix form as

$$\mathbf{Y}_j = \sqrt{\frac{E_1}{N_t}} \mathbf{D}^{(1)} \mathbf{A}_j^{(1)} \mathbf{F} + \mathbf{N}_j^{\text{tot}}, \tag{3.46}$$

in which $\mathbf{F}$ is an $L^{(1)} \times L$ matrix whose (l, l')th element is $f(l, l')$. The multipath gain coefficient matrix $\mathbf{A}_j^{(1)}$ of size $N_t \times L^{(1)}$ is of a form similar to (3.39). Subsequently, the decision rule can be stated as

$$\hat{\mathbf{D}}^{(1)} = \underset{\mathbf{D}^{(1)}}{\operatorname{argmin}} \sum_{j=1}^{N_r} \left\| \mathbf{Y}_j - \sqrt{\frac{E_1}{N_t}} \mathbf{D}^{(1)} \mathbf{A}_j^{(1)} \mathbf{F} \right\|^2 . \tag{3.47}$$

3.6 PERFORMANCE ANALYSIS

In this section we provide performance analysis of single-band UWB systems. We consider a multiuser UWB system with N_u users, each equipped with N_t transmitter and N_r receiver antennas. For the channel model, we consider the case of Nakagami-m fading where the amplitude of the lth path, $|\alpha_{ij}^{(u)}(l)|$, is modeled as a Nakagami-m random variable with average power $E[|\alpha_{ij}^{(u)}(l)|^2] = \Omega_u(l)$. The MIMO channel coefficients are assumed to be real, mutually independent, and quasistatic. For each user, the time delay $\tau^{(u)}(l)$ and the average power $\Omega_u(l)$ are assumed identical for every transmitter–receiver link.

To analyze the performances of three different systems discussed in Section 3.5, we first calculate the noise and interference statistics. From $n_{j,k'}(l')$ defined in (3.34), we have

$$E[n_{j,k'}(l')] = \int_{-\infty}^{\infty} v_{k'}^{(1)}(t - \tau^{(1)}(l'))E[n_j(t)]dt = 0. \tag{3.48}$$

The variance of $n_{j,k'}(l')$ can be computed as

$$E[|n_{j,k'}(l')|^2] = \int_{-\infty}^{\infty}\int_{-\infty}^{\infty} v_{k'}^{(1)}(s - \tau^{(1)}(l'))v_{k'}^{(1)}(t - \tau^{(1)}(l')) \underbrace{E[n_j(s)n_j(t)]}_{N_0/2\delta(s-t)} ds\, dt$$

$$= \frac{N_0}{2}\int_{-\infty}^{\infty} [v_{k'}^{(1)}(t - \tau^{(1)}(l'))]^2 dt \triangleq \sigma_n^2. \tag{3.49}$$

Hence, $n_{j,k'}(l')$ is a zero-mean Gaussian random variable with variance σ_n^2. Next, we investigate the distribution of multiple access interference $n_{j,k'}^{mui}(l')$. Defining

$$n_{i,k'}^{(u)}(l, l') \triangleq \int_{-\infty}^{\infty} v_{k'}^{(1)}(t - \tau^{(1)}(l'))x_i^{(u)}(t - \tau^{(u)}(l))dt, \tag{3.50}$$

$n_{j,k'}^{mui}(l')$ [defined in (3.34)] can be reexpressed as

$$n_{j,k'}^{mui}(l') = \sum_{u=2}^{N_u}\sum_{i=1}^{N_t}\sum_{l=0}^{L-1} \alpha_u^{ij}(l)n_{i,k'}^{(u)}(l, l'). \tag{3.51}$$

Using the same approach as in [Win00], one can show that $n_{i,k'}^{(u)}(l, l')$ is approximately Gaussian random variable with zero mean and variance

$$E[|n_{i,k'}^{(u)}(l, l')|^2] = \frac{E_u}{N_t}\frac{1}{T_f}\int_{-\infty}^{\infty}\left[\int_{-\infty}^{\infty} w(t - s)v(t)\, dt\right]^2 ds.$$

$$\triangleq \frac{E_u}{N_t}\sigma_a^2. \tag{3.52}$$

Assuming independent Nakagami fading coefficients, the statistics of $n_{j,k'}^{mui}(l')$ can be evaluated as follows:

$$E\left[n_{j,k'}^{mui}(l')\right] = \sum_{u=2}^{N_u}\sum_{i=1}^{N_t}\sum_{l=0}^{L-1}E\left[\alpha_u^{ij}(l)\right]E\left[n_{i,k'}^{(u)}(l,l')\right] = 0 \qquad (3.53)$$

and

$$E\left[\left|n_{j,k'}^{mui}(l')\right|^2\right] = \sum_{u=2}^{N_u}\sum_{i=1}^{N_t}\sum_{l=0}^{L-1}\underbrace{E\left[\left|\alpha_u^{ij}(l)\right|^2\right]}_{= \, \Omega_u(l)}\underbrace{E\left[\left|n_{i,k'}^{(u)}(l,l')\right|^2\right]}_{= \, (E_u/N_t)\sigma_a^2}$$

$$= \sigma_a^2\sum_{u=2}^{N_u}E_u\sum_{l=0}^{L-1}\Omega_u(l). \qquad (3.54)$$

Applying the central limit theorem for sufficiently large L, N_t, and N_u, $n_{j,k'}^{mui}(l')$ can be approximated as a Gaussian random variable with zero mean and variance $E[|n_{j,k'}^{mui}(l')|^2]$. Therefore, we can model the total noise and interference $n_{j,k'}^{tot}(l') = n_{j,k'}^{mui}(l') + n_{j,k'}(l')$ as a zero-mean Gaussian random variable with variance

$$E\left[\left|n_{j,k'}^{tot}(l')\right|^2\right] = E\left[\left|n_{j,k'}^{mui}(l')\right|^2\right] + E[|n_{j,k'}(l')|^2]$$

$$= \sigma_a^2\sum_{u=2}^{N_u}E_u\sum_{l=0}^{L-1}\Omega_u(l) + \sigma_n^2$$

$$\stackrel{\Delta}{=}\sigma_{n^{tot}}^2. \qquad (3.55)$$

Since the total noise and interference can be approximated with Gaussian distribution, pairwise error probability (PEP) can be evaluated in a fashion similar to that in a conventional narrowband MIMO system. Such PEP calculation relies on the detection rule, which is different for distinct modulation and multiple access schemes. In addition, since both σ_n^2 and σ_a^2 [defined in (3.49) and (3.52) respectively], are in terms of the reference signal $v(t)$, their values, and hence the statistics of $n_{j,k'}^{tot}(l')$, also depend on particular modulation and multiple access techniques. PEP evaluations for TH-BPPM, TH-BPSK, and DS-BPSK are given in the following subsections. The user superscript, (u), will be omitted to simplify the notation.

3.6.1 TH-BPPM

Recall from (3.15) that the reference signal for a TH-BPPM system is a delayed version of $v(t) = w(t) - w(t - T_d)$. Thus, we have

$$\int_{-\infty}^{\infty}[v(t)]^2dt = \int_{-\infty}^{\infty}[w(t) - w(t - T_d)]^2dt = 2[1 - \gamma(T_d)]. \qquad (3.56)$$

Substituting (3.56) into (3.49), the noise variance is found to be $\sigma_a^2 = [1 - \gamma(T_d)]N_0$. Next, replacing $v(t)$ in (3.52) with $w(t) - w(t - T_d)$, the value of σ_a^2 can be evaluated as

$$\sigma_a^2 = \frac{1}{T_f} \int_{-\infty}^{\infty} \left[\int_{-\infty}^{\infty} w(t-s)[w(t) - w(t-T_d)] \, dt \right]^2 ds$$

$$= \frac{1}{T_f} \int_{-\infty}^{\infty} [\gamma^2(s) + \gamma^2(s-T_d) - 2\gamma(s)\gamma(s-T_d)] \, ds$$

$$\triangleq 2(\bar{\sigma}_a^2 - \sigma_d^2), \tag{3.57}$$

where $\bar{\sigma}_a^2 = 1/T_f \int_{-\infty}^{\infty} \gamma^2(s) ds$ and $\sigma_d^2 = 1/T_f \int_{-\infty}^{\infty} \gamma(s)\gamma(s-T_d) \, ds$. Therefore, using (3.55), the variance of $n_{j,k'}^{\text{tot}}(l')$ is given by

$$\sigma_{n^{\text{tot}}}^2 = 2(\bar{\sigma}_a^2 - \sigma_d^2) \sum_{u=2}^{N_u} E_u \sum_{l=0}^{L-1} \Omega_u(l) + [1 - \gamma(T_d)] N_0. \tag{3.58}$$

Suppose that $\mathbf{D}$ and $\hat{\mathbf{D}}$ are two distinct transmitted space–time codewords. Following calculation steps similar to those in [Pro01], the PEP conditioned on the channel coefficient matrix $\mathbf{A}_j$ is given by

$$P(\mathbf{D} \to \hat{\mathbf{D}} \,|\, \mathbf{A}_j) = Q \left(\sqrt{ \frac{\rho}{2N_t} \sum_{j=1}^{N_r} \|(\mathbf{D} - \hat{\mathbf{D}})\mathbf{A}_j\|^2 } \right), \tag{3.59}$$

where $Q(x)$ is the Gaussian error function, defined as

$$Q(x) = \frac{1}{\sqrt{2\pi}} \int_x^{\infty} \exp\left(-\frac{s^2}{2} \right) ds \tag{3.60}$$

and

$$\rho = \frac{[1 - \gamma(T_d)]^2 E_1}{2\sigma_{n^{\text{tot}}}^2}. \tag{3.61}$$

Substituting (3.58) into (3.61), we obtain

$$\rho = \left[4 \frac{\bar{\sigma}_a^2 - \sigma_d^2}{[1 - \gamma(T_d)]^2} \sum_{u=2}^{N_u} \frac{E_u}{E_1} \sum_{l=0}^{L-1} \Omega_u(l) + \frac{2N_0}{[1 - \gamma(T_d)] E_1} \right]^{-1}. \tag{3.62}$$

Note that if all users have equal transmitted power $E_1 = E_2 = \cdots = E_{N_u} \triangleq E$, (3.62) becomes

$$\rho = \left[4 \frac{\bar{\sigma}_a^2 - \sigma_d^2}{[1 - \gamma(T_d)]^2} \sum_{u=2}^{N_u} \sum_{l=0}^{L-1} \Omega_u(l) + \left(\frac{[1 - \gamma(T_d)] E}{2N_0} \right)^{-1} \right]^{-1}. \tag{3.63}$$

Applying the inequality $Q(x) \leq \frac{1}{2} \exp(-x^2/2)$, for $x > 0$, the conditional PEP in (3.59) can be upper bounded by

$$P(\mathbf{D} \rightarrow \hat{\mathbf{D}} \mid A_j) \leq \frac{1}{2} \exp\left(-\frac{\rho}{4N_t} \sum_{j=1}^{N_r} \|(\mathbf{D} - \hat{\mathbf{D}})A_j\|^2\right). \tag{3.64}$$

For convenience, let us define

$$\mathbf{Z} = (\mathbf{D} - \hat{\mathbf{D}})^{\mathrm{T}}(\mathbf{D} - \hat{\mathbf{D}}), \tag{3.65}$$

where $(\cdot)^{\mathrm{T}}$ denotes transpose operation. The term $\|(\mathbf{D} - \hat{\mathbf{D}})A_j\|^2$ in (3.64) can be expressed as

$$\|(\mathbf{D} - \hat{\mathbf{D}})A_j\|^2 = \sum_{l=0}^{L-1} \mathbf{a}_j^{\mathrm{T}}(l)\mathbf{Z}\mathbf{a}_j(l), \tag{3.66}$$

where $\mathbf{a}_j(l)$ denotes the lth column of A_j. Since $\mathbf{Z}$ is a real symmetric matrix, there exists a set of nonnegative eigenvalues $\{\lambda_i\}_{i=1}^{N_t}$ and the corresponding normalized eigenvectors $\{\mathbf{v}_i\}_{i=1}^{N_t}$ such that

$$\mathbf{Z} = \mathbf{V}\Lambda\mathbf{V}^{\mathrm{T}}, \tag{3.67}$$

where $\mathbf{V} \stackrel{\Delta}{=} [\mathbf{v}_1\mathbf{v}_2\cdots\mathbf{v}_{N_t}]$ is an orthonormal matrix and Λ is a diagonal matrix whose diagonal elements are the eigenvalues of $\mathbf{Z}$. Substituting (3.67) into (3.66), we have

$$\|(\mathbf{D} - \hat{\mathbf{D}})A_j\|^2 = \sum_{l=0}^{L-1} \mathbf{a}_j^{\mathrm{T}}(l)\mathbf{V}\Lambda\mathbf{V}^{\mathrm{T}}\mathbf{a}_j(l) = \sum_{l=0}^{L-1} \sum_{i=1}^{N_t} \lambda_i|\beta_{ij}(l)|^2, \tag{3.68}$$

in which $\beta_{ij}(l) \stackrel{\Delta}{=} \mathbf{a}_j^{\mathrm{T}}(l)\mathbf{v}_i$. Since $\{\alpha_{ij}(l)\}_{i=1}^{N_t}$ are independent identically distributed and $\mathbf{V}$ is orthonormal, $\{\mathbf{v}_0, \mathbf{v}_1, \ldots, \mathbf{v}_{N_t}\}$ is an orthonormal basis of $\mathcal{R}^{N_t}$ and $\{\beta_{ij}(l)\}_{i=1}^{N_t}$ are independent random variables whose magnitude is approximately Nakagami-m distributed with parameter $\tilde{m} = N_t m/(N_t m - m + 1)$ and average power $\Omega_1(l)$ [Nak60, p. 25]. By the use of characteristic functions the PDF of $|\beta_{ij}(l)|^2$ is given by [Sim00]

$$p_{|\beta_{ij}(l)|^2}(x) = \frac{1}{\Gamma(\tilde{m})}\left(\frac{\tilde{m}}{\Omega_1(l)}\right)^{\tilde{m}} x^{\tilde{m}-1} \exp\left(-\frac{\tilde{m}}{\Omega_1(l)}x\right). \tag{3.69}$$

Substituting (3.68) into (3.64) and averaging (3.64) with respect to the distribution of $|\beta_{ij}(l)|^2$, the resulting PEP upper bound can be found as

$$P(\mathbf{D} \rightarrow \hat{\mathbf{D}}) \leq \left[\prod_{l=0}^{L-1}\prod_{i=1}^{N_t}\left(1 + \frac{\rho}{4N_t}\frac{\Omega_1(l)}{\tilde{m}}\lambda_i\right)\right]^{-\tilde{m}N_r}. \tag{3.70}$$

For high-SNR environments, the bound in (3.70) can be simplified further to

$$P(\mathbf{D} \rightarrow \hat{\mathbf{D}}) \leq \left(\prod_{l=0}^{L-1} \prod_{i=1}^{r} \frac{\rho}{4N_t} \frac{\Omega_1(l)}{\tilde{m}} \lambda_i \right)^{-\tilde{m}N_r} = \left[G(\tilde{m}) \frac{\rho}{4N_t} \right]^{-\tilde{m}rN_rL}, \qquad (3.71)$$

where $G(\tilde{m}) \triangleq (\tilde{m})^{-1} (\prod_{l=0}^{L-1} \Omega_1(l))^{1/L} (\prod_{i=1}^{r} \lambda_i)^{1/r}$, r is the rank, and the $\{\lambda_i\}_{i=1}^{r}$ represent nonzero eigenvalues of matrix $\mathbf{Z}$. For a single-user system, since there is no effect of multiple access interference, ρ in (3.63) reduces to $[1 - \gamma(T_d)]E_1/2N_0$. Thus, the PEP upper bound in (3.71) becomes

$$P(\mathbf{D} \rightarrow \hat{\mathbf{D}}) \leq \left[G(\tilde{m}) \frac{[1 - \gamma(T_d)] E_1}{8N_t N_0} \right]^{-\tilde{m}rN_rL}. \qquad (3.72)$$

In this case, the exponent $\tilde{m}rN_rL$ determines the slope of the performance curve plotted as a function of SNR, whereas the product $G(\tilde{m})$ displaces the curve. Hence, the minimum values of $\tilde{m}rN_rL$ and $G(\tilde{m})$ over all pairs of distinct codewords define the diversity gain and coding gain, respectively. Note that $r \leq N_t$; therefore, the maximum achievable diversity gain is $\tilde{m}N_tN_rL$.

3.6.2 TH-BPSK

Since the reference signal for a TH-BPSK system is the shifted monocycle whose energy is unity (i.e., $\int_{-\infty}^{\infty} [v(t)]^2 \, dt = \int_{-\infty}^{\infty} [w(t)]^2 \, dt = 1$), the noise variance becomes $\sigma_n^2 = N_0/2$. In addition, substituting $v(t) = w(t)$ in (3.52), we have $\sigma_a^2 = \bar{\sigma}_a^2$, where $\bar{\sigma}_a^2$ is defined in (3.57). Therefore, $n_{j,k'}^{tot}(l')$ is zero-mean Gaussian random variable with variance

$$\sigma_{n_{tot}}^2 = \bar{\sigma}_a^2 \sum_{u=2}^{N_u} E_u \sum_{l=0}^{L-1} \Omega_u(l) + \frac{N_0}{2}. \qquad (3.73)$$

As a result, following the same calculations as in Section 3.6.1, the upper bound of the PEP can be expressed similar to (3.71) as

$$P(\mathbf{D} \rightarrow \hat{\mathbf{D}}) \leq \left[G(\tilde{m}) \frac{\rho}{4N_t} \right]^{-\tilde{m}rN_rL}, \qquad (3.74)$$

where $G(\tilde{m})$ is of the same form as the one defined in (3.71), and

$$\rho = \frac{E_1}{2\sigma_{n_{tot}}^2} = \left[2\bar{\sigma}_a^2 \sum_{u=2}^{N_u} \frac{E_u}{E_1} \sum_{l=0}^{L-1} \Omega_u(l) + \left(\frac{E_1}{N_0} \right)^{-1} \right]^{-1}, \qquad (3.75)$$

which becomes E_1/N_0 for the single-user system.

3.6.3 DS-BPSK

With the spreading sequence $\{c(n_c)\} \in \{-1, 1\}$ being independent and identically distributed (iid) discrete uniform random variables, the variance of $n_{k'}^j(l')$ can be found from (3.17) and (3.49) as

$$
\sigma_n^2 = \frac{N_0}{2} \frac{1}{N_c} \int_{-\infty}^{\infty} \mathrm{E} \left[\sum_{n_c'=0}^{N_c-1} c(n_c') w(t - k'T_f - n_c'T_c) \right]^2 dt
$$

$$
= \frac{N_0}{2} \frac{1}{N_c} \sum_{n_c'=0}^{N_c-1} \int_{-\infty}^{\infty} [w(t - k'T_f - n_c'T_c)]^2 \, dt = \frac{N_0}{2}. \tag{3.76}
$$

Substituting (3.17) into (3.52) results in

$$
\sigma_a^2 = \frac{1}{T_f} \int_{-\infty}^{\infty} \left[\sqrt{\frac{1}{N_c}} \sum_{n_c'=0}^{N_c-1} c(n_c') \int_{-\infty}^{\infty} w(t - s) w(t - n_c'T_c) \, dt \right]^2 ds.
$$

$$
= \frac{1}{T_f} \frac{1}{N_c} \sum_{n_c'=0}^{N_c-1} \int_{-\infty}^{\infty} \gamma^2(s) \, ds = \bar{\sigma}_a^2. \tag{3.77}
$$

Observe that both σ_n^2 and σ_a^2 for DS-BPSK have the same values as those for TH-BPSK. Hence, the variance of $n_{j,k}^{\text{tot}}(l')$ can be expressed similar to (3.73). As with the case of TH-BPSK, the upper bound of the PEP conditioned on the channel matrix is given by

$$
P(\mathbf{D} \to \hat{\mathbf{D}} \,|\, \mathbf{A}_j) \leq \frac{1}{2} \exp \left(-\frac{\rho}{4N_t} \sum_{j=1}^{N_r} \|(\mathbf{D} - \hat{\mathbf{D}}) \mathbf{A}_j \mathbf{F}\|^2 \right), \tag{3.78}
$$

where $\rho = E_1/2\sigma_{n^{\text{tot}}}^2$ is in the same form as (3.75). The term $\|(\mathbf{D} - \hat{\mathbf{D}}) \mathbf{A}_j \mathbf{F}\|^2$ can be evaluated as follows:

$$
\|(\mathbf{D} - \hat{\mathbf{D}}) \mathbf{A}_j \mathbf{F}\|^2 = \mathrm{tr} \left(\mathbf{F}^T \mathbf{A}_j^T \mathbf{Z} \mathbf{A}_j \mathbf{F} \right), \tag{3.79}
$$

where $\mathbf{Z}$ is defined in (3.65) and $\mathrm{tr}(\cdot)$ denotes the trace operation [i.e., $\mathrm{tr}(\mathbf{X})$ is the sum of diagonal elements of $\mathbf{X}$]. By the use of eigenvalue decomposition as in (3.67), we have

$$
\|(\mathbf{D} - \hat{\mathbf{D}}) \mathbf{A}_j \mathbf{F}\|^2 = \mathrm{tr} \left(\mathbf{F}^T \mathbf{A}_j^T \mathbf{V} \mathbf{\Lambda} \mathbf{V}^T \mathbf{A}_j \mathbf{F} \right) = \mathrm{tr} \left(\mathbf{F}^T \mathbf{B}_j^T \mathbf{\Lambda} \mathbf{B}_j \mathbf{F} \right), \tag{3.80}
$$

in which $\mathbf{B}_j \triangleq \mathbf{v}^T \mathbf{A}_j$ is an $N_t \times L_0$ matrix whose (i, l)th element is $\beta_{ij}(l)$, as defined in the TH-UWB case. Since $\mathbf{V}$ is an orthonormal matrix and the elements of $\mathbf{A}_j$ at each column are independent identically Nakagami-m distributed, the $\{\beta_{ij}(l)\}$ are independent. Define $\tilde{\mathbf{B}}_j = \mathbf{B}_j \mathbf{F}$ and denote its (i, l')th element by $\tilde{\beta}_{ij}(l')$ [i.e.,

$\tilde{\beta}_{ij}(l') = \sum_{l=0}^{L^{(l)}-1} \beta_{ij}(l) f(l, l')]$. We can simplify (3.80) to

$$\|(\mathbf{D} - \hat{\mathbf{D}})\mathbf{A}_j\mathbf{F}\|^2 = \text{tr}\left(\tilde{\mathbf{B}}_j^{\mathsf{T}}\Lambda^{\hat{}}\tilde{\mathbf{B}}_j\right) = \sum_{l=0}^{L-1}\sum_{i=1}^{N_t}\lambda_i|\tilde{\beta}_{ij}(l)|^2. \tag{3.81}$$

Thus, the conditioned PEP upper bound in (3.78) can be expressed as

$$P(\mathbf{D} \to \hat{\mathbf{D}} \mid \mathbf{A}_j) \le \frac{1}{2}\exp\left(-\frac{\rho}{4N_t}\sum_{j=1}^{N_r}\sum_{l=0}^{L-1}\sum_{i=1}^{N_t}\lambda_i|\tilde{\beta}_{ij}(l)|^2\right). \tag{3.82}$$

The PEP upper bound can be found by averaging (3.82) with respect to the joint distribution of $\{|\tilde{\beta}_{ij}(l)|^2\}$. Since the $\{|\tilde{\beta}_{ij}(l)|\}_{l=0}^{L-1}$ are correlated Nakagami random variables, their joint distribution is difficult to obtain. Therefore, instead of evaluating the average PEP upper bound directly, we quantify the performance merits of DS-UWB space–time system by investigating the term $\sum_{j=1}^{N_r}\|(\mathbf{D} - \hat{\mathbf{D}})\mathbf{A}_j\mathbf{F}\|^2$ as follows.

Let us first define $\Delta = \mathbf{I}_{N_rL} \otimes \Lambda$ where $\mathbf{I}_x$ is the identity matrix of size $x \times x$ and $\otimes$ denotes the tensor product. Denote the vector operation as $\text{vec}(\mathbf{X}) = [\mathbf{x}_1^{\mathsf{T}}\mathbf{x}_2^{\mathsf{T}}\cdots\mathbf{x}_N^{\mathsf{T}}]^{\mathsf{T}}$, in which $\mathbf{x}_n$ is the nth column of $\mathbf{X}$, and define a column vector

$$\tilde{\mathbf{b}} \triangleq \left[(\text{vec}(\tilde{\mathbf{B}}_1))^{\mathsf{T}}(\text{vec}(\tilde{\mathbf{B}}_2))^{\mathsf{T}}\cdots\left(\text{vec}(\tilde{\mathbf{B}}_{N_r})\right)^{\mathsf{T}}\right]^{\mathsf{T}} \tag{3.83}$$

of length N_tN_rL. Then it follows from (3.81) that

$$\sum_{j=1}^{N_r}\|(\mathbf{D} - \hat{\mathbf{D}})\mathbf{A}_j\mathbf{F}\|^2 = \sum_{j=1}^{N_r}\text{tr}((\tilde{\mathbf{B}}_j)^{\mathsf{T}}\Lambda\tilde{\mathbf{B}}_j) = \tilde{\mathbf{b}}^{\mathsf{T}}\Delta\tilde{\mathbf{b}}. \tag{3.84}$$

Now we can rewrite (3.78) as

$$P(\mathbf{D} \to \hat{\mathbf{D}} \mid \mathbf{A}_j) \le \frac{1}{2}\exp(-\frac{\rho}{4N_t}\tilde{\mathbf{b}}^{\mathsf{T}}\Delta\tilde{\mathbf{b}}). \tag{3.85}$$

Let $\mathbf{R} = \text{E}[\tilde{\mathbf{b}}\tilde{\mathbf{b}}^{\mathsf{T}}]$ denote the correlation matrix of $\tilde{\mathbf{b}}$. Since the correlation matrix is nonnegative definite, it has a symmetric square root $\mathbf{U}$ such that $\mathbf{R} = \mathbf{U}^{\mathsf{T}}\mathbf{U}$ [Hor85]. Let $\mathbf{q} = (\mathbf{U}^{\mathsf{T}})^{-1}\tilde{\mathbf{b}}$. Since the correlation matrix of $\mathbf{q}$ is

$$\text{E}[\mathbf{q}\mathbf{q}^{\mathsf{T}}] = \text{E}[(\mathbf{U}^{\mathsf{T}})^{-1}\tilde{\mathbf{b}}\tilde{\mathbf{b}}^{\mathsf{T}}\mathbf{U}^{-1}] = (\mathbf{U}^{\mathsf{T}})^{-1}\mathbf{R}\mathbf{U}^{-1} = \mathbf{I}_{N_tN_rL}, \tag{3.86}$$

the components of $\mathbf{q}$ are uncorrelated. Substituting $\tilde{\mathbf{b}} = \mathbf{U}^{\mathsf{T}}\mathbf{q}$ into (3.85), we arrive at

$$P(\mathbf{D} \to \hat{\mathbf{D}} \mid \mathbf{A}_j) \le \frac{1}{2}\exp\left(-\frac{\rho}{4N_t}\mathbf{q}^{\mathsf{T}}\mathbf{U}\Delta\mathbf{U}^{\mathsf{T}}\mathbf{q}\right). \tag{3.87}$$

Assuming that $\mathbf{R}$ is full rank, $\mathbf{U}$ is also full rank [Hor85]. Therefore, with the same argument as in the case of TH-UWB, by replacing $\mathbf{Z}$ with $\mathbf{U}\Delta\mathbf{U}^{\mathsf{T}}$ it follows that maximum diversity gain can be achieved by maximizing the rank of Δ. Note that

$$\text{rank}(\Delta) = N_rL\,\text{rank}(\Lambda) = N_rL\,\text{rank}(\mathbf{Z}), \tag{3.88}$$

where rank(**X**) stands for the rank of **X**. Hence, the rank criterion for a DS-UWB space–time system is identical to that of a TH-UWB space–time system. That is, the diversity gain can be maximized when **Z** is of full rank. To quantify the coding gain, it might be necessary to evaluate the statistics of **q**, which is difficult to obtain for the Nakagami fading distribution. In the following section we perform simulations to investigate further the performance of a DS-UWB space–time system.

3.7 SIMULATION RESULTS

In this section, simulated performance of single-band UWB systems is described. The simulations are performed for TH-BPPM, TH-BPSK, and DS-BPSK systems. We employ UWB signals with a frame interval $T_f = 100$ ns and a pulse duration T_w of 0.8 ns. The pulse transmitted is a Gaussian monocycle. To accommodate the effect of a propagation channel and the variation of antenna characteristics caused by a large frequency bandwidth, the monocycle received is modeled as the second derivative of the Gaussian pulse [Win00]:

$$w(t) = \sqrt{(8/3)}(1 - 4\eta)\exp(-2\eta), \tag{3.89}$$

where $\eta = \pi(t/\tau_o)^2$ and τ_o ($\tau_o \approx 0.4T_w$) parameterizes the width of the monocycle. The factor $\sqrt{8/3}$ is introduced such that each monocycle has unit energy. The autocorrelation function of the pulse in (3.89) is given by [Ram98]

$$\gamma(t) = (1 - 4\eta + (4/3)\eta^2)\exp(-\eta). \tag{3.90}$$

The second derivative of the Gaussian monocycle and its normalized autocorrelation function are shown in Fig. 3.13(a) and (b), respectively. Note that for a single-user system, utilizing a rectangular monocycle whose autocorrelation function is zero for $|t| \geq T_w$ yields the same performance as using the second derivative of a Gaussian pulse. Therefore, $\gamma(t)$ in (3.90) can be approximately zero for $|t| \geq T_w$.

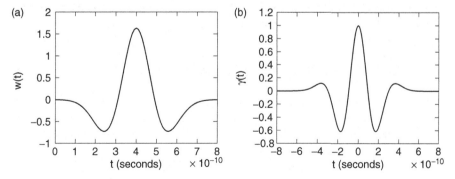

Figure 3.13 (a) Second derivative of the Gaussian monocycle waveform used at the receiver; (b) its autocorrelation function.

The data transmitted are binary symbols taking a value from $\{-1, 1\}$ with equal probability. In the TH-BPPM system, the modulation delay $T_d = \text{argmin}_{T_d} \int_{-\infty}^{\infty} w(t)w(t - T_d) = 0.22T_w$. Since an interval of $T_w + T_d$ seconds is required for one symbol modulation, we choose the hop duration $T_c = T_w + T_d$ seconds. Unlike TH-BPPM, a TH/DS-BPSK system does not require an additional time delay for data modulation, and hence its modulation interval can be made equal to the pulse duration. Therefore, the hop periods for both TH-BPSK and DS-BPSK systems are selected to be $T_c = T_w$. To avoid ISI, the total hop interval is limited to $N_c T_c \leq T_f - \max_u\{\tau_u(L^{(u)} - 1)\}$. Note that with a fixed T_f, since the hop duration of BPPM is larger than that of BPSK, BPPM can support fewer hops than can the BPSK system. To evaluate the performance of an asynchronous system regardless of the choice of the particular code, TH sequence and DS spreading are selected randomly [Som02, Dur03a]. Here, the TH sequence takes values from $\{0, 1, \ldots, N_c - 1\}$ equally, whereas the spreading sequence $\{c(n_c)\}_{n_c=0}^{N_c-1} \in \{-1, 1\}$ with equal probability.

We employ a frequency-selective channel model in which the delay profile is generated according to [Tar03], and the path amplitude is Nakagami-m distributed with $m = 2$. The power of the $L^{(u)}$ paths are normalized such that $\sum_{l=0}^{L^{(u)}-1} \Omega_u(l) = 1$. We assume that the power delay profiles of all users are similar. The channels are quasistatic over a K symbol period. The channel coefficients, transmitted signals, and noise are generated independently. Unless specified otherwise, the number of fingers for the RAKE receiver is fixed to be $L = 4$.

We adopt the real orthogonal design (ROD) [Tar99] as the space–time code for the MIMO UWB system. For simplicity, we focus on the MIMO UWB system employing two transmit antennas. Generalization to MIMO UWB systems with a higher number of transmitter antennas is straightforward. Based on ROD structure, the space–time code is modeled as

$$\mathbf{D} = \begin{pmatrix} d_1 & d_2 \\ -d_2 & d_1 \end{pmatrix}. \tag{3.91}$$

Since two time slots are used for two-symbol (d_0 and d_1) transmission, the code is of full rate $R = 1$. The corresponding matrix $\mathbf{Z}$ is given by

$$\mathbf{Z} = \sum_{i=1}^{2}(d_i - \hat{d}_i)^2 \mathbf{I}_2 = 4\sum_{i=1}^{2} \delta(d_i - \hat{d}_i)\mathbf{I}_2. \tag{3.92}$$

For an ROD code with rate $1/K$, where $K \geq 2$ is an even integer, a codeword $\mathbf{D}$ of size $K \times 2$ is modeled as

$$\mathbf{D} = d\left(\begin{pmatrix} 1 & 1 \\ -1 & 1 \end{pmatrix}^T \cdots \begin{pmatrix} 1 & 1 \\ -1 & 1 \end{pmatrix}^T\right)^T_{K \times 2}. \tag{3.93}$$

In this case, the data symbol d is transmitted repeatedly over K frames from both transmitter antennas. The matrix $\mathbf{Z}$ becomes

$$\mathbf{Z} = (d - \hat{d})^2 K \mathbf{I}_2 = 4K\delta(d - \hat{d})\mathbf{I}_2. \tag{3.94}$$

Observe that both full- and reduced-rate codes result in two equal eigenvalues $\lambda_0 = \lambda_1 \overset{\Delta}{=} \lambda$ and the matrix $\mathbf{Z} = \lambda \mathbf{I}_2$ of full rank ($r = 2$). Substituting the eigenvalues into (3.71), we arrive at

$$P(\mathbf{D} \to \hat{\mathbf{D}}) \le \left[\prod_{l=0}^{L-1} \left(\frac{\Omega(l)}{\tilde{m}} \frac{\rho\lambda}{8} \right) \right]^{-2\tilde{m}N_r}, \tag{3.95}$$

where $\tilde{m} = 2m/(m+1)$. With channel parameters and N_r being fixed, (3.95) depends only on the value of $\rho\lambda$. The higher the $\rho\lambda$, the better the performance.

To compare the performance of various systems, we assume that the energy per bit E_b is fixed. For simplicity, we also assume that all users have equal energy transmitted per frame (E). Expressing E in terms of E_b, we have $E = E_b$ for full-rate and $E = E_b/K$ for $1/K$-rate codes. Assuming one erroneous symbol, the eigenvalues for the $1/K$ rate are K times larger than those for a full-rate code. We denote the eigenvalue of the full-rate code as $\bar{\lambda}$.

Figure 3.14(a) and (b) show the BER performance of TH and DS UWB systems in single- and multiuser environments. We can see from both figures that MIMO systems outperform SISO systems, and increasing the number of receiver antennas yields better performance regardless of the modulation and multiple access techniques. Consider the single-user case illustrated in Fig. 3.14(a). At any fixed SNR, TH-BPSK and DS-BPSK systems are similar in performance, and both BPSK systems yield performance superior to that of TH-BPPM. In Fig. 3.14(b) we show system performance when five asynchronous users are active. In the low-SNR regime, simulation results are similar to those of the single-user case. That is, TH-BPSK and DS-BPSK outperform the TH-BPPM scheme, and both BPSK systems yield similar performances. However, due to the multiple access interference, the BER of TH multiuser systems drop slightly with increasing E_b/N_0, and a high error floor can be noticed at high SNR. This is due to the fact that in the high-SNR regime, it is the effect of multiuser interference that prevails regardless of the E_b/N_0. We can also observe from Fig. 3.14(b) that the performance of TH-BPSK degrades faster than that of TH-BPPM. This means that at high SNR, a TH-BPSK system is more vulnerable than a TH-BPPM system to multiple access interference. On the other hand, even in multiple access scenarios, we can still see considerable improvement in a DS-BPSK space–time system. Therefore, it is evident that among the analyzed schemes, a DS-BPSK space–time system provides the best performance in multiple access environments.

In Fig. 3.15(a) and (b), we plot the BER performances as a function of the number of active users with a fixed E_b/N_0 of 4 and 12 dB, respectively. In both cases we observe performance degradation when more users are presented. For any number of users, BPSK systems achieve better performance than a BPPM system. In comparisions

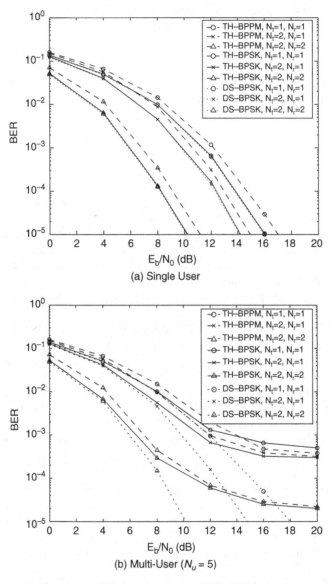

Figure 3.14 Performance of TH and DS UWB systems.

between the TH and DS techniques, DS-BPSK performs slightly better than TH-BPSK at $E_b/N_0 = 4$ dB, and it outperforms TH-BPSK remarkably at $E_b/N_0 = 12$ dB, especially when an ROD space–time system is utilized. This is because with a fixed T_f and random spreading sequence, the multiple access interference of DS systems is less than that of TH systems [Bou03, Som02]. As we can see from Fig. 3.15(b), when the number of users increases, the BER of a TH space–time system increases much faster than that a DS space–time system. Therefore, we can conclude that

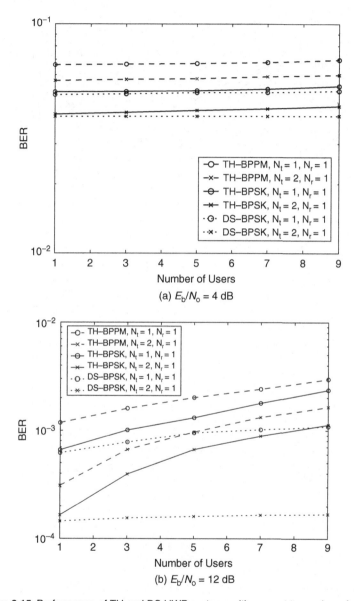

Figure 3.15 Performance of TH and DS UWB systems with respect to number of users.

a DS space–time system is capable of accommodating multiple users with a lower BER than that of TH systems.

Figure 3.16(a) and (b) demonstrate the effect of RAKE fingers to the performances of TH and DS schemes. Here we consider UWB space–time systems with two transmitter and one receiver antennas. The BER versus E_b/N_0 curves for single- and multiuser systems, each employing RAKE receivers with $L = 1$, 4 and 8 fingers, are

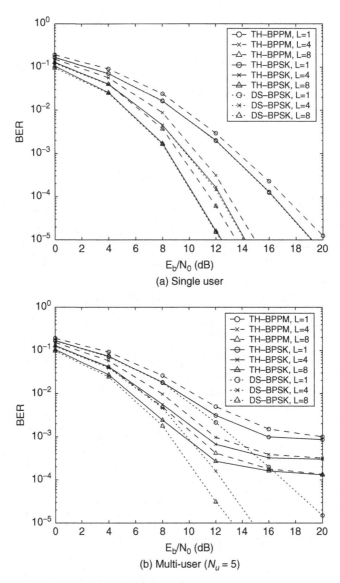

Figure 3.16 Performance of TH and DS UWB-MIMO systems with various RAKE fingers, *L*.

shown in Fig. 3.16(a) and (b), respectively. The performance improvement with the increasing number of fingers can be observed from the figure. This corresponds to the fact that a RAKE receiver with more fingers provides increased capability to capture the available signal energy in dense multipath environments. Such improvements can obviously be seen in the single-user case. This supports our analytical results in previous sections that the diversity gain is increasing with *L*. Nevertheless, in

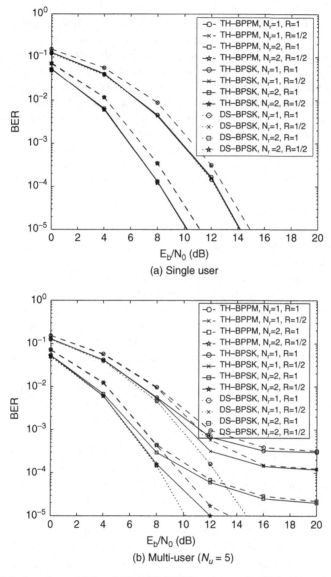

Figure 3.17 Performance of TH and DS UWB-MIMO systems with ROD space–time codes of different rates.

the presence of multiple access interference, the performance improvement of TH systems degrades rapidly, as shown in Fig. 3.16(b). On the contrary, the benefit of additional fingers is evident for a DS-BPSK system in both single- and multiuser scenarios.

In Fig. 3.17(a) and (b), we show the performance of single- and multiple-user systems employing ROD space–time codes with full and half rates. Both figures illustrate that utilizing either a full- or reduced-rate code, BPSK provides less BER than does the BPPM scheme. From Fig. 3.17(a) we can see that the performances of full- and half-rate ROD codes are close to each other for every modulation schemes. Unlike the single-user case, the results in Fig. 3.17(b) show that when the code rate is lower, both TH-BPSK and TH-BPPM multiuser systems achieve better performances, especially in the high-SNR regime. However, for DS-BPSK multiple access systems, the BER improvement obtained from reducing the code rate is insignificant. This is because for a DS multiuser system with $N_u = 5$, the effect of multiple access interference is quite small, and a ROD space–time code provides close to maximum achievable performance without decreasing the code rate. Once again, DS-BPSK outperforms other modulation schemes for both full and half rates.

3.8 CHAPTER SUMMARY

In this chapter we present UWB systems employing single-band approaches. We describe various modulation and multiple access techniques as well as receiver processing for single-band UWB systems. We consider both single- and multiantenna UWB systems. We also provide performance analysis for multiuser single-band UWB systems with various transmission schemes, including TH-BPPM, TH-BPSK, and DS-BPSK. The BER performance is analyzed under Nakagami-m frequency-selective fading channels. For space–time-coded UWB systems, the performance metrics (diversity and coding gains) are quantified regardless of the particular coding scheme. We show that the use of space–time coding in combination with RAKE architecture is able to exploit spatial diversity as well as multipath diversity inherent in UWB environments. An example of UWB space–time signals based on a ROD space–time code for two-transmitter-antenna systems is considered. Comparing various modulation techniques, we show that in the single-user case, the performance improvement using MIMO transmission is more significant for TH-BPSK and DS-BPSK than for TH-BPPM, whereas in multiple access scenarios, DS-BPSK outperforms other schemes. For example, by employing two transmitter and one receiver antennas for a system of five users and $E_b/N_0 = 8$ dB, the BER for TH-BPPM decreases from 1.5×10^{-2} to 9.7×10^{-3}, for TH-BPSK from 10^{-2} to 5.6×10^{-3}, and for DS-BPSK from 10^{-2} to 4.6×10^{-3}. In addition, we show that reducing the rate of UWB space–time code would not improve the performance of single-user systems for all modulation schemes. However, in multiuser environments, reducing the code rate improves the performances of TH systems, while the improvement in DS systems is not significant.

4

UWB: MULTIBAND OFDM APPROACH

A variety of UWB systems can be designed to use the available UWB spectrum of 7.5 GHz. The traditional design approach is based on a sequence of impulselike waveforms that occupies a very wide spectrum, as discussed in Chapter 3. Although the impulse architectures offer relatively simple radio designs, they provide little flexibility in spectrum management. Moreover, building RF and analog circuits as well as high-speed analog-to-digital converters to process this ultrashort pulse signal is a challenging problem. In addition, the digital complexity needs to be quite large (e.g., a high number of RAKE fingers) in order to capture sufficient multipath energy to meet the range requirements.

Multiband approaches were proposed [Sab03, Foe03a, Bat03, Bat04] in which the UWB frequency band is divided into several subbands. Each subband occupies a bandwidth of at least 500 MHz in compliance with FCC regulations. By interleaving the symbols across subbands, multiband UWB can maintain the transmitter power as if a large GHz bandwidth were being utilized. The advantage is that the multiband approach allows the information to be processed over a much smaller bandwidth, thereby reducing overall design complexity as well as improving spectral flexibility and worldwide compliance. To capture the multipath energy efficiently, the orthogonal frequency-division multiplexing (OFDM) technique has been used to modulate the information in each subband. The multiband OFDM approach was one of the two leading proposals for the IEEE 802.15.3a wireless personal area networking (WPAN) standard [TG3a] in 2004 and has been approved as the UWB standard by the European Computer Manufacturers Association (ECMA) [ECM05] in December 2005.

In this chapter we first present the fundamental concept of the multiband OFDM approach. Next, we describe the physical layer design of multiband OFDM. We provide a baseband implementation of the multiband OFDM approach following the multiband OFDM proposal in the IEEE 802.15.3a standard. The baseband transceiver design, including the scrambler, channel encoder, interleaver, and OFDM modulator,

Ultra-Wideband Communications Systems: Multiband OFDM Approach, By W. Pam Siriwongpairat and K. J. Ray Liu
Copyright © 2008 John Wiley & Sons, Inc.

is presented in detail. Finally, we present the media access control (MAC) layer design of the multiband OFDM system according to the IEEE 802.15.3 WPANs standard.

In Section 4.1 we present an overview of the multiband OFDM approach, including fundamental concepts and the signal model. In Section 4.2 we describe the multiband OFDM approach that has been proposed in the IEEE 802.15.3a standard. The physical design of a multiband OFDM system is presented in Section 4.3, and the MAC layer design is described in Section 4.4.

4.1 OVERVIEW OF MULTIBAND OFDM APPROACH

4.1.1 Fundamental Concepts

The principal idea of multiband approaches is to divide the UWB frequency band (3.1 to 10.6 GHz) into multiple smaller frequency bands (also referred to as *subbands*) and uses multiple carrier frequencies to transmit the information. Each subband has a bandwidth greater than 500 MHz to comply with the FCC definition of a UWB signal. Figure 4.1 shows a multiband UWB spectrum. In this example, each UWB signal occupies 500 MHz of bandwidth, and eight of them cover the total bandwidth of 4 GHz. Each of these signals can be transmitted simultaneously to achieve a very high bit rate. The signals can also be interleaved across subbands to maintain the transmitter

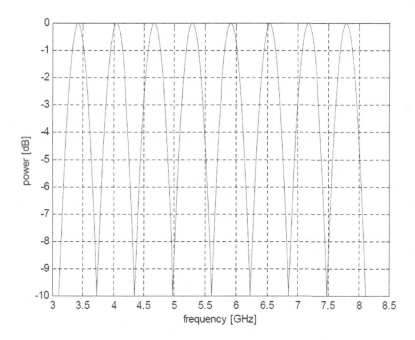

Figure 4.1 Multiband spectrum [Dis03].

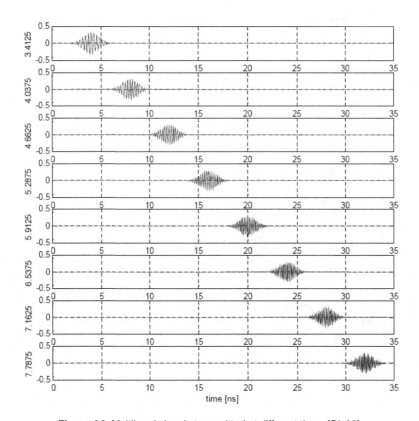

Figure 4.2 Multiband signals transmitted at different times [Dis03].

power as if the large GHz bandwidth is used while allowing multiple users to transmit at the same time. Figure 4.2 shows a time-domain representation of multiband UWB signals in which the signals at different center frequencies are staggered in time and transmitted at different discrete times. In the figure the center frequencies of the signals relative to the individual bands are shown in the vertical axis [Dis03].

With the multiband approach, the information can now be processed over a much smaller bandwidth, thereby reducing overall design complexity as well as improving spectral flexibility and worldwide compliance. Specifically, the smaller bandwidth eases the requirement on the sampling rate of an analog-to-digital converter. Also, it increases the ability for more fine-grained control of the power spectral density such that the average transmitter power can be maximized while still meeting the FCC spectral mask. The multiband approach also enables the UWB system to comply with local regulations by dynamically turning off certain tones or channels in software.

To efficiently capture the multipath energy, which is richly inherent in a UWB environment, the OFDM technique has been used to modulate the information in each subband. The major difference between multiband OFDM and traditional OFDM schemes is that the multiband OFDM symbols are not sent continually on one frequency band; instead, they are interleaved over different subbands across both time and frequency. Multiple access of multiband UWB is enabled by the use of suitably designed frequency-hopping sequences over the set of subbands.

4.1.2 Signal Model

Consider a UWB multiband OFDM system with the available UWB spectrum divided into S subbands. Each subband occupies a bandwidth $BW > 500$ MHz and the OFDM has N subcarriers, as shown in Fig. 4.3. At each OFDM symbol period, the modulated symbol is transmitted over one of the S subbands. These symbols are time-interleaved across subbands.

Let $d_k(n)$ denote the complex coefficient to be transmitted in subcarrier n during the kth OFDM symbol period. The coefficient $d_k(n)$ can consist of data symbols, pilots, or training symbols. The baseband signal is constructed in similarly to a conventional OFDM system, as shown in Fig. 4.4. In particular, each OFDM symbol $x_k(t)$ is constructed using an inverse Fourier transform:

$$x_k(t) = \sum_{n=0}^{N-1} d_k(n) \exp(\mathbf{j}2\pi n \, \Delta f t), \tag{4.1}$$

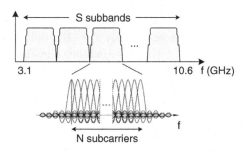

Figure 4.3 UWB multiband OFDM spectrum.

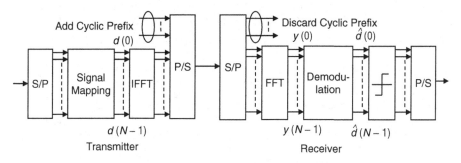

Figure 4.4 Transmitter and receiver of an OFDM system.

where $\Delta f = $ BW/N is the frequency spacing between the adjacent subcarriers. The resulting waveform has a duration of $T_{\text{FFT}} = 1/\Delta f$. The cyclic prefix of length T_{CP} is appended in order to mitigate the effects of multipath interference and to transform the multipath linear convolution into a circular convolution. Also, the guard interval of length T_{GI} is added at the end of the OFDM block. The guard interval is used to provide more flexibility in the implementation. For instance, it can be used to provide sufficient time for switching between bands, to relax the analog transmitter and receiver filters, to relax filter specifications for adjacent channel rejection, or to help reduce the peak-to-average power ratio (PARP). The symbol duration becomes $T_{\text{SYM}} = T_{\text{FFT}} + T_{\text{CP}} + T_{\text{GI}}$. The complex baseband signal $x_k(t)$ is modulated to the RF signal with carrier frequency f_k. The RF signal transmitted can be modeled as

$$s(t) = \sum_k \text{Re}\{x_k(t - kT_{\text{SYM}})\exp(j2\pi f_k t)\}. \qquad (4.2)$$

The carrier frequency, f_k, specifies the subband, in which the signal is transmitted during the kth OFDM symbol duration. These carrier frequency sequences are based on time–frequency codes, which are assigned uniquely to various users so as to minimize the multiple access interference.

4.2 IEEE 802.15.3A WPAN STANDARD PROPOSAL

In the multiband OFDM proposal for IEEE 802.15.3a WPAN standard [TG3a], the UWB signal is shaped so that it occupies only 528 MHz of bandwidth, allowing 14 such signals to cover the entire 7.5-GHz band, as shown in Fig. 4.5. Each 528-MHz band uses OFDM, which allows each UWB band to be divided into a set of orthogonal narrowband channels (with a much longer symbol period duration).

4.2.1 OFDM Parameters

The OFDM has 128 subcarriers, and quadrature phase shift keying (QPSK) is used to modulate the transmitter signal at the subcarriers. The reason for using QPSK is due to the limitation of the transmitter power, which is not allowed to exceed −41.3 dBm/MHz [FCC02]. Each OFDM symbol is preappended with a

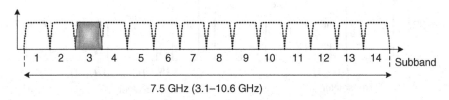

Figure 4.5 Band division in multiband OFDM standard proposal.

TABLE 4.1   Timing Parameters

Parameters	Value
Number of OFDM subcarriers	128
Number of data subcarriers	100
Number of defined pilot subcarriers	12
Number of guard subcarriers	10
Δf: Subcarrier frequency spacing	4.125 MHz ($=528$ MHz/128)
T_{FFT}: IFFT/FFT period	242.42 ns ($1/\Delta f$)
T_{CP}: Cyclic prefix duration	60.61 ns ($=32/528$ MHz)
T_{GI}: Guard interval duration	9.47 ns ($=5/528$ MHz)
T_{SYM}: Symbol duration	312.5 ns ($T_{FFT} + T_{CP} + T_{GI}$)

zero-trailing prefix to eliminate the ISI. The guard interval is appended to the OFDM symbol to ensure a transition between the two consecutive symbols. The timing parameters associated with the OFDM symbols [Bat03] are summarized in Table 4.1.

4.2.2 Rate-Dependent Parameters

The multiband OFDM system can support 10 different data rates, including 53.3, 55, 80, 106.7, 110, 160, 200, 320, 400, and 480 Mbps. The modulation is based on QPSK for every data rate. Different data rates are achieved by using a different channel coding rate, frequency spreading gain, or time spreading gain [Bat04]. The channel coding rate can be 1/3, 11/32, 1/2, 5/8, or 3/4. The frequency spreading gain can be either 1 or 2. The frequency spreading gain of 2 is obtained by choosing conjugate symmetric inputs to the IFFT. The time-domain spreading gain can also be either 1 or 2. The time-domain spreading gain of 2 is achieved by repeating the same information in an OFDM symbol on two different subbands.

The channel coding rate, frequency spreading gain, and time spreading gain for different data rates are summarized in Table 4.2. The overall spreading gain, shown in the last column in Table 4.2, is the multiplication of time- and frequency-domain spreading gain.

Figure 4.6 illustrates the time and frequency spreading for UWB signal with different data rates. For rates not higher than 80 Mbps, both time and frequency spreadings are performed, yielding an overall spreading gain of 4. In other words, the same information is repeated four times, as depicted in Fig. 4.6(a). For rates between 106.7 and 200 Mbps, only time-domain spreading is utilized, which results in an overall spreading gain of 2. As shown in Fig. 4.6(b), all subcarriers are used to transmit different information (i.e., the information is not repeated in the frequency domain) and the information is transmitted twice over two OFDM symbol periods. A system with data rates higher than 200 Mbps exploits neither frequency nor time spreading, and the overall spreading gain is 1. As illustrated in Fig. 4.6(c), the subcarriers and time slots are used to transmit different information.

TABLE 4.2 Rate-Dependent Parameters

Data Rate (Mbps)	Modulation	Coding Rate, R	Frequency Spreading Gain	Time Spreading Gain	Overall Spreading Gain	Coded Bits per Symbol (N_{CBPS})
53.3	QPSK	1/3	2	2	4	100
55	QPSK	11/32	2	2	4	100
80	QPSK	1/2	2	2	4	100
106.7	QPSK	1/3	1	2	2	200
110	QPSK	11/32	1	2	2	200
160	QPSK	1/2	1	2	2	200
200	QPSK	5/8	1	2	2	200
320	QPSK	1/2	1	1	1	200
400	QPSK	5/8	1	1	1	200
480	QPSK	3/4	1	1	1	200

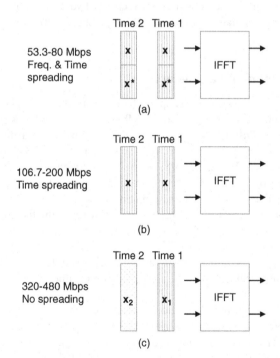

Figure 4.6 Time–frequency spreading: (a) low rates; (b) middle rates; (c) high rates.

4.2.3 Operating Band Frequencies

The center frequency of UWB signal in each 528-MHz band is given by [Bat03]

$$f_c(\text{MHz}) = 2904 + 528 n_b, \quad n_b = 1, 2, \ldots, 14. \tag{4.3}$$

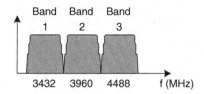

Figure 4.7 Operating band frequencies in mandatory mode.

It was argued in [Bat03] that due to path loss, attenuation, and losses in current RF technology, the use of frequency bands above 5 GHz yields only minor improvements in capacity at the cost of increased receiver complexity. For this reason it was suggested that only three bands be located below the 5-GHz band as a mandatory mode. The frequency operation in this mode is depicted in Fig. 4.7.

4.2.4 Channelization

The channelization in multiband OFDM system is based on a set of time–frequency codes. Each code specifies the center frequency for the transmission of each OFDM symbol. For example, Fig. 4.8 illustrates a time–frequency representation of multiband OFDM signal with time–frequency code {1 3 2 1 3 2}. The time–frequency codes are used to provide frequency diversity as well as to enable simultaneous piconet operation with little multiple access interference.

In a multiband OFDM proposal for the IEEE 802.15.3a standard, a set of four time–frequency codes are proposed, as shown in Table 4.3. These codes are designed such that the average number of collisions between any two codes is 1/3.

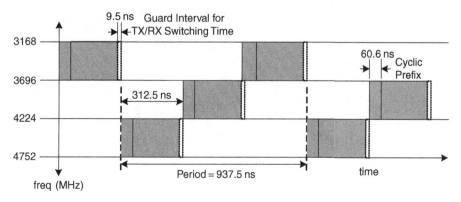

Figure 4.8 Time–frequency representation of multiband OFDM signal with time–frequency code {1 3 2 1 3 2} [Bat04]. (Copyright © 2004 IEEE.)

TABLE 4.3 Time–Frequency Codes for Different Piconets [Bat04] (Copyright © 2004 IEEE)

Piconet Number	Time–Frequency Codes					
1	1	2	3	1	2	3
2	1	3	2	1	3	2
3	1	1	2	2	3	3
4	1	1	3	3	2	2

4.3 PHYSICAL LAYER DESIGN

In this section we describe the physical layer design of a multiband OFDM system as proposed in the IEEE 802.15.3a standard. The structure of a multiband OFDM transmitter and receiver is shown in Fig. 4.9(a). At the transmitter, the data bit stream is first scrambled, encoded with the convolutional code, interleaved, mapped into a sequence of QPSK samples, and OFDM-modulated. The resulting OFDM baseband signal is up-converted to a specified subband, passed through a power amplifier (PA), and finally, transmitted. A different pattern of band switching is assigned to different users in order to gain frequency diversity while minimizing the multiple access interference.

The corresponding receiver structure is shown in Fig. 4.9(b). The RF signal received is passed through a preselect filter, amplified with a low-noise amplifier (LNA), down-converted to baseband, scaled in amplitude by a voltage gain amplifier (VGA), and then digitized. Next, synchronization is performed, and QPSK symbols are demodulated from the OFDM baseband signal. From the QPSK symbols, an estimated

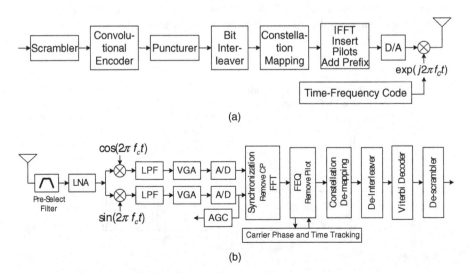

(a)

(b)

Figure 4.9 Multiband OFDM system (a) transmitter; (b) receiver [Bat04]. (Copyright © 2004 IEEE.)

transmitted bit sequence is reconstructed and de-interleaved. Then the sequence is channel-decoded using a Viterbi decoder. Finally, the decoder output is descrambled, yielding an estimated sequence of the bit information transmitted.

In the following sections we describe each functional block in the multiband OFDM transceiver in details.

4.3.1 Scrambler and De-scrambler

The first block in the baseband of the UWB transmitter is a data scrambler. The purpose of the data scrambler is to convert a data bit sequence into a pseudorandom sequence that is free from long strings of simple patterns such as marks and spaces. The polynomial generator of the pseudorandom binary sequence (PRBS) is

$$g(D) = 1 + D^{14} + D^{15}, \tag{4.4}$$

where D represents a single bit delay. The corresponding PRBS x_n is generated as

$$x_n = x_{n-14} \oplus x_{n-15}, \tag{4.5}$$

where "$\oplus$" represents modulo-2 addition. The scrambled data bit stream is obtained as

$$s_n = b_n \oplus x_n, \tag{4.6}$$

where b_n is the unscrambled data bit stream.

The scrambler and de-scrambler are initialized with the same *seed value*, which is chosen based on the first two bits, b_0 and b_1, in the unscrambled data sequence. They are called the *seed identifiers*. The correspondence between seed values and seed identifiers follows Table 4.4.

4.3.2 Convolutional Encoder and Viterbi Decoder

The second block in the baseband of the UWB transmitter is the convolutional encoder and puncturer. This block serves to add patterns of redundancy to the data in order to improve the SNR for more accurate decoding at the receiver. The system supports five different coding rates: 1/3, 11/32, 1/2, 5/8, and 3/4. The generator polynomials are $g_0 = [133_8]$, $g_1 = [145_8]$, and $g_2 = [175_8]$ for the code corresponding to coding rate 1/3, called the *mother code* and the constraint length is $K = 7$. Figure 4.10

TABLE 4.4   Scrambler Seed Selection [Bat03]

Seed Identifier (b_1, b_0)	Seed Value $(x_{14} \cdots x_0)$
0,0	0011 11111111 111
0,1	0111 11111111 111
1,0	1011 11111111 111
1,1	1111 11111111 111

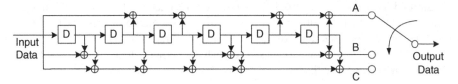

Figure 4.10 Convolutional encoder (mother code) [Bat03].

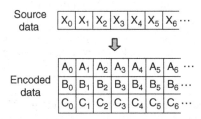

Figure 4.11 Input–output relation of convolutional encoder [Bat03].

illustrates the mother convolutional encoder. An input data bit produces 3 output bits and thus yields the coding rate 1/3. These output bits are denoted as A, B, and C, where A is the first bit, B is the second bit, and C is the last bit. They are formed in ABC order to yield the output sequence. Figure 4.11 shows the input–output relation of the mother convolutional code. At the receiver, a Viterbi decoder is employed to decode the convolutional encoded sequence.

Other coding rates are obtained through puncturing the mother code. The puncturer simplifies implementation since we need not have other convolutional encoders for these coding rates. Puncturing is a procedure for omitting some encoded bits at the transmitter and inserting a dummy "zero" metric into the sequence received at the receiver in place of the bits omitted. The dummy zero metric inserted should result in no change to the error accumulated in the decoding process [Fle06]. Figure 4.12 describes how coding rate 11/32 can be obtained from the mother code. Eleven bits from x_0 to x_{10} are input to the encoder and result in 33 output bits from A_0 to C_{10}. By omitting bit C_{10}, we obtain a coding rate of 11/32 (i.e., that 11 input bits produce 32 output bits). At the receiver we insert a dummy bit in the place C_{10} and then input the sequence received to a Viterbi decoder to obtain 11 decoded bits.

In a similar manner, we also obtain other coding rates. Figures 4.13, 4.14, and 4.15 show the puncturing patterns of the coding rates 1/2, 5/8, and 3/4, respectively, and illustrate the puncturing process of these coding rates.

4.3.3 Bit Interleaver and De-interleaver

The third block in the baseband of the UWB transmitter is the bit interleaver. The purpose of the bit interleaver is to provide robustness against burst errors. Bit interleaving

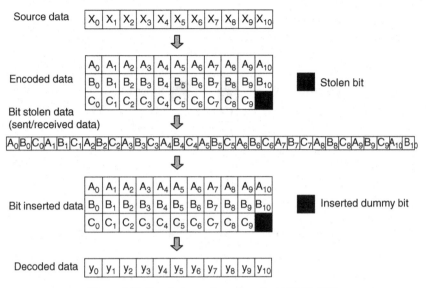

Figure 4.12 Puncturing patterns for rate 11/32 [Bat03].

operates through two stages, a symbol interleaving followed by a tone interleaving. In *symbol interleaving*, the bits across six OFDM symbols are permuted to exploit frequency diversity across the subbands. In *tone interleaving*, the bits across data tones within an OFDM symbol are permuted to exploit frequency diversity across tones. The input–output relations in the symbol interleaving and tone interleaving are as follows [Bat03]:

$$S(i) = U \left\{ \text{Floor} \left(\frac{i}{N_{\text{CBPS}}} \right) + 6 \, \text{Mod}(i, N_{\text{CBPS}}) \right\} \qquad (4.7)$$

$$T(i) = S \left\{ \text{Floor} \left(\frac{i}{N_{\text{Tint}}} \right) + 10 \, \text{Mod}(i, N_{\text{Tint}}) \right\}, \qquad (4.8)$$

where $U(i)$ is the input of the symbol interleaver; $S(i)$ is the output of the symbol interleaver and hence the input of the tone interleaver; $T(i)$ is the output of the tone interleaver; the index $i = 0, \ldots, 6N_{\text{CBPS}} - 1$; Floor($\cdot$) and Mod($\cdot$) denote the floor and modulo functions, respectively; N_{CBPS}, the number of bits per OFDM symbol, follows Table 4.2; and $N_{\text{Tint}} = N_{\text{CBPS}}/10$.

Figure 4.16 illustrates the bit interleaving operation. Six hundred bits $x_0, x_1, \ldots, x_{599}$, equivalent to six OFDM symbols, are input to the symbol interleaver that produces the output in the pattern $x_0, x_6, \ldots x_{594}, x_1, x_7, \ldots, x_{595}, \ldots, x_5, x_{11}, \ldots, x_{599}$. These output bits, now denoted as $y_0, y_1, \ldots, y_{599}$, are input to the tone interleaver that produces the output in the pattern $y_0, y_{10}, \ldots y_{90}, y_1, y_{11}, \ldots, y_{91}, \ldots, y_{509}, y_{519}, \ldots, y_{599}$.

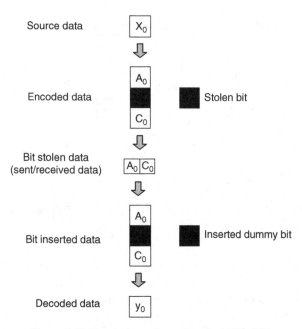

Figure 4.13 Puncturing patterns for rate 1/2 [Bat03].

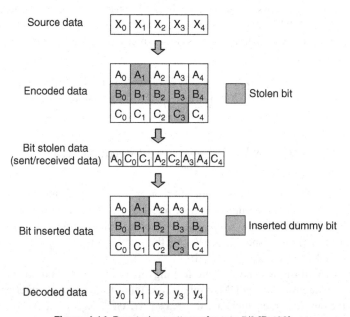

Figure 4.14 Puncturing patterns for rate 5/8 [Bat03].

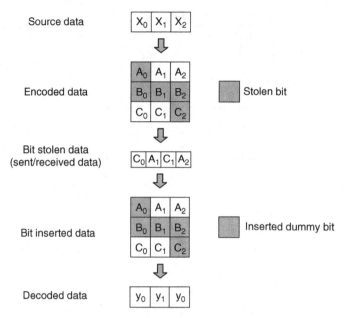

Figure 4.15 Puncturing patterns for rate 3/4 [Bat03].

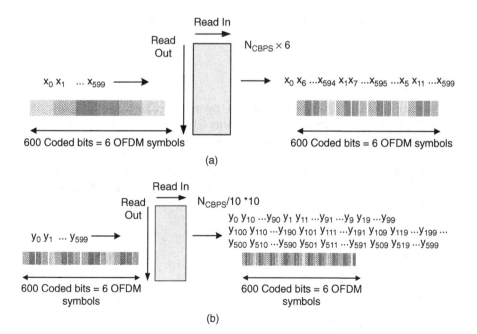

Figure 4.16 Interleaver: (a) symbol interleaver; (b) tone Interleaver [Bat03].

TABLE 4.5   QPSK Encoding [Bat03]

Input Bits (b_1, b_0)	I_{out}	Q_{out}
0,0	−1	−1
0,1	−1	1
1,0	1	−1
1,1	1	1

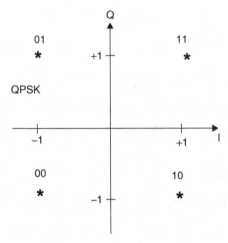

Figure 4.17 QPSK constellation bit encoding [Bat03].

4.3.4 Constellation Mapper

The fourth block in the baseband of the UWB transmitter is the constellation mapper, in which OFDM subcarriers are modulated using QPSK modulation. An input binary sequence is now converted into a complex-valued sequence according to Gray-coded constellation mapping as shown in Fig. 4.17. Based on a pair of input bits, we determine the in-phase and quadrature values, denoted as I and Q, respectively. Their relation follows Table 4.5. From these I and Q values, complex-valued sequences d are obtained as

$$d = (I + jQ)K_{MOD}, \tag{4.9}$$

where $j = \sqrt{-1}$ is an imaginary number and $K_{MOD} = 1/\sqrt{2}$ normalizes d to unit energy.

4.3.5 OFDM Modulation

The complex-valued sequence generated from the constellation mapper is ready for OFDM modulation. The sequence in series is now converted to parallel, and the

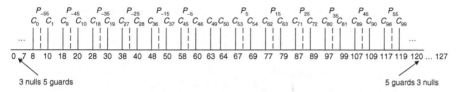

Figure 4.18 Subcarrier frequency allocation [Bat03].

pilots, guards, and nulls are also inserted to the OFDM symbols before IFFT is taken. Each OFDM symbol contains 128 subcarriers. The duration for the OFDM symbol is $T_s = 242.42$ ns. After that, the cyclic prefix used to eliminate the ISI is preappended to the OFDM symbol and the guard interval is used to ensure that a smooth transition between two adjacent OFDM symbols is appended. The cyclic prefix and the guard interval are filled with zeros (such a cyclic prefix is called *zero trailing*). The duration of the cyclic prefix is $T_c = 60.61$ ns, equivalent to 32 subcarriers. The duration of the guard interval is $T_g = 9.47$ ns, equivalent to five subcarriers.

Among the 128 subcarriers of the OFDM symbol, 100 data tones are used to transmit information. Twelve pilot tones are used to ensure the coherent detection robust against frequency offset and phase noise. Ten guard tones are used for a number of purposes, including relaxing the specifications on transmitting and receiving filters. There are also six null tones. The positions of these tones are shown in Fig. 4.18.

For modes with data rates of 80 Mbps or lower, the complex-valued sequence is divided into groups of 50 complex numbers. The data tones c_n in Fig. 4.18 for the kth OFDM symbol relate to the complex-valued sequence d as

$$c_{n,k} = d_{n+50k}$$
$$c_{(n+50),k} = d^*_{(49-n)+50k},$$

(4.10)

where $n = 0, 1, \ldots, 49$, and $k = 0, 1, \ldots, N_{\text{SYM}} - 1$, with N_{SYM} denoting the number of OFDM symbols. Since the same information is transmitted twice using two subcarriers, we obtain the frequency diversity with a spreading gain factor of 2 for these modes.

For modes with data rates larger than 80 Mbps, the complex-valued sequence is divided into groups of 100 complex numbers. The data tones c_n in Fig. 4.18 for the kth OFDM symbol relate to the complex-valued sequence d as

$$c_{n,k} = d_{n+100k},$$

(4.11)

where $n = 0, 1, \ldots, 99$ and $k = 0, 1, \ldots, N_{\text{SYM}} - 1$. In these modes we have no frequency-diversity gain since each subcarrier conveys different information.

Figure 4.18 also shows the positions for the pilots, denoted as P_n, where

$$P_n = \begin{cases} \dfrac{1+j}{\sqrt{2}} & n = 15, 45 \\[2mm] \dfrac{-1-j}{\sqrt{2}} & n = 5, 25, 35, 55 \\[2mm] 0 & \text{otherwise.} \end{cases} \qquad (4.12)$$

For modes with data rates less than 106.67 Mbps, $P_{n,k} = P^*_{-n,k}$, and for modes with data rates 106.67 Mbps or higher, $P_{n,k} = P_{-n,k}$, where $n = -5, -15, -25, -35, -45,$ and -55. The P_n is further BPSK-modulated by a pseudorandom binary sequence p_l to prevent the generation of spectral lines, where $p_{0...127} = \{1, 1, 1, 1, -1, -1, -1,$
$1, -1, -1, -1, -1, 1, 1, -1, 1, -1, -1, 1, 1, -1, 1, 1, -1, 1, 1, 1, 1, 1, 1, -1, 1, 1,$
$1, -1, 1, 1, -1, -1, 1, 1, 1, -1, 1, -1, -1, -1, 1, -1, 1, -1, -1, 1, -1, -1, 1, 1,$
$1, 1, 1, -1, -1, 1, 1, -1, -1, 1, -1, 1, -1, 1, 1, -1, -1, -1, 1, 1, -1, -1, -1, -1,$
$1, -1, -1, 1, -1, 1, 1, 1, 1, -1, 1, -1, 1, -1, 1, -1, -1, -1, -1, -1, 1, -1, 1, 1,$
$-1, 1, -1, 1, 1, 1, -1, -1, 1, -1, -1, -1, 1, 1, 1, -1, -1, -1, -1, -1, -1, -1\}$.
The 10 guards in the kth OFDM symbol take values of

$$P_{n,k} = p_{\text{mod}(k+l,127)} \frac{1+j}{\sqrt{2}} \qquad \text{for } l = 0, 1, 2, 3, 4 \quad \text{and} \quad n = 57 + l, \quad (4.13)$$

where p_l is the pseudorandom binary sequence above. For modes with data rates less than 106.67 Mbps, $P_{n,k} = P^*_{-n,k}$, and for modes with data rates 106.67 Mbps or higher, $P_{n,k} = P_{-n,k}$, where $n = -57, ..., -61$.

4.4 MAC LAYER DESIGN

In this section we present the UWB MAC layer design based on the MAC layer specification in the IEEE 802.15.3 standard [TG3], which is designed to support ad hoc networking and provide multimedia capabilities. The network topology, frame architecture, and network operations under the IEEE 802.15.3 MAC layer protocol are described.

4.4.1 Network Topology

The UWB devices in WPAN are communicating on a centralized network topology called a *piconet*. A piconet comprises a master device, called a *piconet coordinator* (PNC), and multiple slave devices (DEVs) associated with the master device. The PNC is responsible for the task of maintaining piconet operation as follows. First, the PNC periodically transmits beacons that carry necessary information for piconet operations. Such beacon frames enable new devices to join the piconet. Second, the PNC manages the quality of service (QoS) and power-save modes. Finally, the PNC allocates resources for channel access to other devices in the piconet.

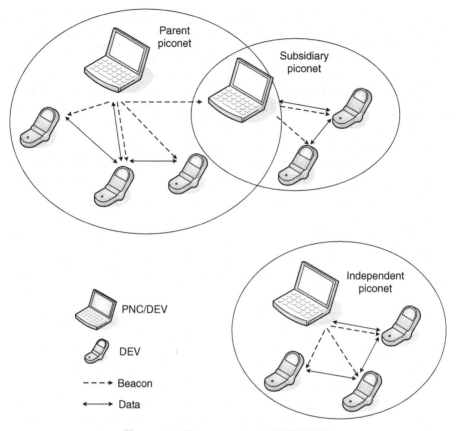

Figure 4.19 Piconet topology in UWB WPAN.

Any UWB device can act as a PNC or a DEV. Typically, the first UWB device that sends a beacon becomes the PNC for the piconet. Control can also be handed off later to another device in the piconet. Once a device joins the piconet, it can transmit data to other devices in the piconet during its assigned time slots. This allows a device to communicate not only to the PNC but also to communicate directly with other devices in the piconet.

Figure 4.19 illustrates the IEEE802.15.3 piconet topology. The figure shows three piconets: the independent, parent, and subsidiary piconets. An *independent piconet* is a piconet that is located far apart or operates on different frequency channels from other piconets. The independent piconets operate independent of one another.

A *parent piconet* is a piconet that allocates time slots to subsidiary piconets in the same system, whereas a *subsidiary piconet* is a piconet that is dependent on the parent piconet. Specifically, a subsidiary piconet is formed under an established piconet, wherein the established piconet becomes the parent piconet. The subsidiary piconet requires a time allocation in the parent piconet and is synchronized with the timing of the parent piconet. Parent and subsidiary piconets share a common frequency channel. The overlapping coverage between the subsidiary and parent piconets can vary from congruent to mostly nonoverlapping with the parent coverage area. There are generally two types of subsidiary piconets:

- *Child piconet*: a subsidiary piconet in which the PNC is a member of the parent piconet. The PNC in a child piconet can exchange information with the PNC in the parent piconet.
- *Neighbor piconet*: a subsidiary piconet in which the PNC is not a member of the parent piconet. The PNC in the neighbor piconet only shares the frequency channel with the parent piconet, but not exchanging any information with the PNC in the parent piconet.

4.4.2 Frame Architecture

Frame architecture in the IEEE 802.15.3 MAC protocol is based on the notion of a piconet superframe, as depicted in Fig. 4.20. A superframe comprises three major parts as follows:

- *Beacon*: transmitted by the PNC at the beginning of the superframe. The beacon provides control and management information to the entire piconet. It also sets timing allocations for the current superframe. The beacon allows DEVs in the piconet to synchronize to the piconet and to know piconet information (e.g., piconet identifier, superframe duration, and channel allocations).
- *Contention access period* (CAP): can be used to transmit signaling or command messages, short data frames, or asynchronous data frames. The channel access during CAP is based on carrier-sense multiple access/collision avoidance (CSMA/CA) with a short request to send and clear-to-send (RTS/CTS) messages. The duration of CAP is specified by the PNC and sent to DEVs in the piconet via the beacon frame. The CAP is optional, and it can be replaced by management CTA slots (as described below).
- *Contention-free period* (CFP): allows each device in the piconet to transmit data to its destination without collision. The channel access during a CFP is based

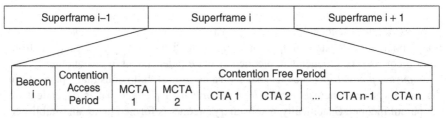

Figure 4.20 Superframe structure for UWB WPAN specified in the IEEE 802.15.3 standard [TG3].

on time-division multiple access (TDMA). In particular, the CFP is divided into multiple time slots called *channel time allocation* (CTA) *slots*. All CTA slots are managed by the PNC through the beacon frame. The CTA slots can be used for communication between the DEVs and the PNC; in this case the slots are called *management CTA* (MCTA) *slots*. The duration of a CFP is also managed by the PNC, and it may vary according to traffic demands.

Each frame transmission in the piconet is followed by an acknowledgment frame. The IEEE 802.15.3 standard offers three modes of acknowledgments: no acknowledgment, immediate acknowledgment, and delayed acknowledgment. If the intended recipient receives the frame correctly, it acknowledges the reception with an immediate acknowledgment. On the other hand, if the frame is not received correctly, no acknowledgment will be transmitted, which corresponds to the no acknowledgment mode. The delay acknowledgment mode is used for directed data frames. In this mode the intended recipient of the directed data frames is allowed to group the acknowledgment indications into a retransmission request command.

4.4.3 Network Operations

In what follows we describe network operations in IEEE 802.15.3 piconets.

- *Starting a piconet.* To create a piconet, a device uses passive scanning to detect existing piconets and then chooses the channel that no existing piconet is using. Once an empty channel is chosen, the device broadcasts its beacon frames in the channel selected. The device then becomes the PNC of the piconet.
- *Joining a piconet.* When a device wants to join a piconet, it uses passive scanning to detect existing piconets and then chooses the channel in which a piconet is established. After selecting the channel, the device authenticates and exchanges capability information with the PNC. Then the device sends an association request to the PNC, and the PNC replies with the association response. Once the device is associated with the piconet, the PNC broadcasts the device information in the beacon frame.
- *Transmitting data in a piconet.* When a device has data to be sent in its associated piconet, the device sends a CTA request to the PNC. The PNC decides whether to satisfy the request according to the resources available. If resources are available, the PNC allocates CTA slots to the device and informs the device via the beacon frame. The device can then transmit its data in the time slots assigned.
- *Leaving a piconet.* A device can leave its associated piconet at any time by sending a disassociation request to the PNC.
- *PNC handover.* When the PNC decides to stop its piconet (e.g., leaves the piconet or runs out of battery power), it can hand over piconet control to another device in the piconet. The PNC selects the best device among those with PNC capability. If no PNC-capable device is found, the PNC simply stops broadcasting

the beacon frames. A PNC handover can also be operated when a new device with more capability than the current PNC joins the piconet. In this case, the PNC can hand over piconet control to the new device. Note that all time-slot allocations are maintained while PNC handover is operating. This allows data to be delivered without interruption during PNC handover.

4.5 CHAPTER SUMMARY

In this chapter we introduce the fundamental concepts of a multiband OFDM approach. First, the system and signal models of the multiband OFDM approach are presented. Then we provide a detailed physical layer design of the multiband OFDM approach proposed in the IEEE 802.15.3a standard. Finally, we present the MAC layer design, which is based on the IEEE 802.15.3 WPAN standard.

5

MIMO MULTIBAND OFDM

To enhance the data rates and transmission ranges of UWB systems, applying the MIMO scheme has attracted considerable interest. In conventional RF technology, MIMO has been well known for its effectiveness in improving system performance in fading environments. Most UWB applications are in a rich scattering indoor environment, which provides an ideal transmission scenario for MIMO implementation. In addition, the GHz center frequency of UWB radio relaxes the requirements on the spacing between antenna array elements. Consequently, the combination of UWB and MIMO technology will become a viable and cost-efficient method to achieve the very high data rate requirements necessary for future short-range wireless applications. In [Yan02, Kum02, Wei03, Sir04, Sir05a], the UWB-MIMO technology has been investigated for traditional single-band UWB systems. Such single-band UWB-MIMO systems are described in Chapter 3. UWB-MIMO-based multiband OFDM systems have recently been proposed and analyzed [Sir06a].

In this chapter we present a general framework to characterize the performance of UWB-MIMO systems with multiband OFDM [Sir05c, Sir06a]. A combination of STF coding and hopping multiband UWB transmission is introduced to exploit all available spatial and frequency diversities. In the performance evaluation, we do not impose any restriction on the delays or the average powers of the multipath components, and the framework presented is applicable for any channel models. Since Nakagami-m statistics can be used to model a wide range of fading conditions, we evaluate the theoretical performances of UWB systems by using the tap-delay-line Nakagami-m fading model, as it can provide some insightful understanding of UWB systems [Foe02a, Sir05a, Sir06a, Foe02a, Fen04]. We quantify the average pairwise error probability as well as the diversity and the coding advantages, regardless of specific coding schemes. As shown in this chapter, the maximum achievable diversity of the MIMO multiband OFDM system is the product of the number of transmitting and receiving antennas, the number of multipath components, and the number of jointly encoded OFDM symbols. An interesting result is that the diversity advantage does

Ultra-Wideband Communications Systems: Multiband OFDM Approach, By W. Pam Siriwongpairat and K. J. Ray Liu
Copyright © 2008 John Wiley & Sons, Inc.

Adapted with permission from © 2006 *IEEE Transactions on Signal Processing*, Vol. 54, no. 1, Jan. 2006, pp. 214–224.

not depend on the fading parameter m. The diversity gain obtained under Nakagami fading with an arbitrary m parameter is almost the same as that obtained in Rayleigh fading, which is equivalent to Nakagami-m fading with $m = 1$.

The principal concept of MIMO-OFDM communications is presented in Section 5.1. In Section 5.2 we present the MIMO multiband OFDM system model, including the signal modulation, channel model, receiver description, and detection technique. The performance analysis of a peer-to-peer MIMO multiband OFDM system is presented in Section 5.3. In Section 5.4 simulation results are presented to support the theoretical analysis.

5.1 MIMO-OFDM COMMUNICATIONS

A MIMO communications system employing multiple transmitting and receiving antennas is known as an efficient technology to improve system performance in fading environments. With a proper MIMO coding design, a MIMO system is able to exploit multipath propagation and hence reduce the detrimental effects of channel fading.

As discussed in Section 3.5, space–time-coded MIMO systems have been designed for frequency-nonselective fading channels. When the fading channel is frequency selective, space–frequency (SF)-coded MIMO-OFDM systems have been shown to be an efficient approach to achieving the benefits of spatial and frequency diversity [Agr98, Blu01, Bol00, Su03, Su05a, Su04a]. Recently, space–time–frequency (STF) codes have also been proposed for MIMO-OFDM systems [Gon01, Mol02, Liu02, Su05c]. By utilizing some proper STF-coding and modulation techniques, STF-coded MIMO systems can exploit all of the spatial, temporal, and frequency diversity and hence promise to yield high spectral efficiency and remarkable performance improvement.

Figure 5.1 illustrates a point-to-point MIMO-OFDM system. With SF coding, the information symbols are jointly encoded across transmitting antennas and OFDM subcarriers. Each SF codeword is transmitted within one OFDM symbol duration. The STF encoder, on the other hand, jointly encodes information across transmitting antennas, OFDM subcarriers, and multiple OFDM blocks.

Consider an SF-coded MIMO-OFDM system with N_t transmitting antennas, N_r receiving antennas, and N OFDM subcarriers. At the transmitter, the information bit

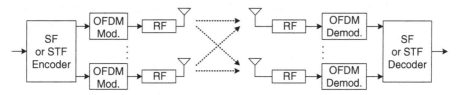

Figure 5.1 MIMO-OFDM communications system.

sequence is mapped into an $N \times N_t$ SF codeword $\mathbf{D}$:

$$\mathbf{D} = \begin{pmatrix} d_1(0) & d_2(0) & \cdots & d_{N_t}(0) \\ d_1(1) & d_2(1) & \cdots & d_{N_t}(1) \\ \vdots & \vdots & \ddots & \vdots \\ d_1(N-1) & d_2(N-1) & \cdots & d_{N_t}(N-1) \end{pmatrix}, \tag{5.1}$$

where $d_i(n)$ represents a complex symbol to be transmitted in subcarrier n at transmitting antenna i. The SF codeword is assumed to satisfy the energy constraint $E[\|\mathbf{D}\|^2] = NN_t$.

Let $H_{ij}(n)$ denote the channel frequency response at subcarrier n in an i–j transmitter–receiver link. The MIMO channels are assumed independent for all transmitter–receiver links. Then the signal received at receiver antenna j in subcarrier n can be expressed as

$$y_j(n) = \sqrt{\frac{E}{N_t}} \sum_{i=1}^{N_t} d_i(n)H_{ij}(n) + z_j(n), \tag{5.2}$$

where E is the average signal energy and $z_j(n)$ is the additive noise sample, which can be modeled as a complex Gaussian random variable with zero mean and a two-sided power spectral density of $N_0/2$.

Assuming that channel-state information is available at the receiver, a maximum-likelihood detection can be performed jointly across all N_r receiver antennas. Specifically, the detector estimated the transmitted symbols according to the detection rule:

$$\hat{\mathbf{D}} = \underset{\mathbf{D}}{\operatorname{argmin}} \sum_{j=1}^{N_r} \sum_{n=1}^{N} \left| y_j(n) - \sqrt{\frac{E}{N_t}} \sum_{i=1}^{N_t} d_i(n)H_{ij}(n) \right|^2. \tag{5.3}$$

Suppose that $\mathbf{D}$ and $\hat{\mathbf{D}}$ are two different SF codeword matrices. Then the pairwise error probability between $\mathbf{D}$ and $\hat{\mathbf{D}}$ can be upper bounded as [Su03]

$$P(\mathbf{D} \to \hat{\mathbf{D}}) \leq \binom{2PN_r - 1}{PN_r} \left(\prod_{p=1}^{P} \lambda_p \right)^{-N_r} \left(\frac{\rho}{N_t} \right)^{-PN_r}, \tag{5.4}$$

where $\rho = E/N_0$, P is the rank of $\Delta \circ \mathbf{R}$, $\mathbf{R}$ is a channel correlation matrix, and $\Delta \triangleq (\mathbf{D} - \hat{\mathbf{D}})(\mathbf{D} - \hat{\mathbf{D}})$, in which $(\cdot)^{\mathcal{H}}$ denotes conjugate transpose operation. The channel correlation matrix R is of size N by N, which is determined as

$$\mathbf{R} = E[\mathbf{HH}^{\mathcal{H}}], \tag{5.5}$$

where

$$\mathbf{H} = [H_{ij}(0)H_{ij}(1) \cdots H_{ij}(N-1)]^{\mathrm{T}}. \tag{5.6}$$

Note that the correlation matrix is the same for any i and j [Su03]. The result in (5.4) indicates that the diversity gain of the SF-coded MIMO-OFDM system is the minimum rank of $\triangle \circ \mathbf{R}$. By looking at the rank of the matrices $\triangle$ and $\mathbf{R}$, one can find that the maximum achievable diversity gain is at most

$$G_d = \min\{L N_t N_r, N N_r\}, \tag{5.7}$$

where L is the number of multipath components.

5.2 MIMO MULTIBAND OFDM SYSTEM MODEL

In this section we present a system model of UWB MIMO multiband OFDM. We consider the UWB multiband OFDM system with a fast band-hopping rate: that is, the signal is transmitted on a frequency band during one OFDM symbol interval, then moved to a different frequency band at the next interval.

5.2.1 Transmitter Description

We consider a peer-to-peer UWB multiband OFDM system with N_t transmitting and N_r receiving antennas, as shown in Fig. 5.2. The information is encoded across N_t transmitting antennas, N OFDM subcarriers, and K OFDM blocks.

At the transmitter, the coded information sequence from a channel encoder is partitioned into blocks of N_b bits. Each block is mapped onto a $KN \times N_t$ STF codeword matrix:

$$\mathbf{D} = \left[\mathbf{D}_0^T \ \mathbf{D}_1^T \ \cdots \ \mathbf{D}_{K-1}^T \right]^T, \tag{5.8}$$

where

$$\mathbf{D}_k = \left[\mathbf{d}_1^k \ \mathbf{d}_2^k \ \cdots \ \mathbf{d}_{N_t}^k \right], \tag{5.9}$$

in which $\mathbf{d}_i^k = [d_i^k(0) \ d_i^k(1) \cdots d_i^k(N-1)]^T$ for $i = 1, 2, \ldots, N_t$ and $k = 0, 1, \ldots, K - 1$. The symbol $d_i^k(n)$, $n = 0, 1, \ldots, N - 1$, represents the complex symbol to be transmitted over subcarrier n by transmitting antenna i during the kth OFDM symbol period. The matrix $\mathbf{D}$ is normalized to have average energy $E[\|\mathbf{D}\|^2] = KNN_t$.

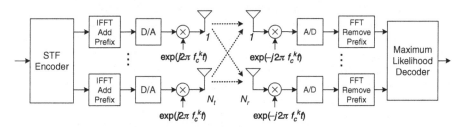

Figure 5.2 MIMO multiband OFDM system.

At the kth OFDM block, the transmitter applies N-point IFFT over each column of the matrix $\mathbf{D}_k$, yielding an OFDM symbol of length T_{FFT}. To mitigate the effect of intersymbol interference, a cyclic prefix of length T_{CP} is added to the output of the IFFT processor.

After adding the cyclic prefix and guard interval, the OFDM symbol is passed through a digital-to-analog converter, resulting in an analog baseband OFDM signal of duration $T_{SYM} = T_{FFT} + T_{CP} + T_{GI}$. The baseband OFDM signal to be transmitted by the ith transmit antenna at the kth OFDM block can be expressed as

$$x_i^k(t) = \sqrt{\frac{E}{N_t}} \sum_{n=0}^{N-1} d_i^k(n) \exp\{(j2\pi n \Delta f)(t - T_{CP})\}, \qquad (5.10)$$

where $t \in [T_{CP}, T_{FFT} + T_{CP}]$, $j \triangleq \sqrt{-1}$, and $\Delta f = 1/T_{FFT} = BW/N$ is the frequency separation between two adjacent subcarriers. The factor $\sqrt{E/N_t}$ guarantees that the average energy per transmitted symbol is E, independent of the number of transmit antennas. In the interval $[0, T_{CP}]$, $x_i^k(t)$ is a copy of the last part of the OFDM symbol, and $x_i^k(t)$ is zero in the interval $[T_{FFT} + T_{CP}, T_{SYM}]$.

The complex baseband signal $x_i^k(t)$ is filtered, up-converted to an RF signal with a carrier frequency f_c^k, and finally sent from the ith transmitting antenna. The multiband OFDM signal transmitted at the transmitting antenna i over K OFDM symbol periods is given by

$$s_i(t) = \sum_{k=0}^{K-1} \text{Re}\{x_i^k(t - kT_{SYM}) \exp(j2\pi f_c^k t)\}. \qquad (5.11)$$

The carrier frequency f_c^k specifies the subband, in which the signal is transmitted during the kth OFDM symbol period. The carrier frequency can be changed from one OFDM block to another, so as to enable frequency diversity while minimizing multiple access interference. The band-hopping rate depends on the channel environment and the data rates desired. Since the signals from all transmit antennas share the same subband, f_c^k is identical for every transmitting antenna. Note that transmissions from all N_t transmitting antennas are simultaneous and synchronous. Figure 5.3

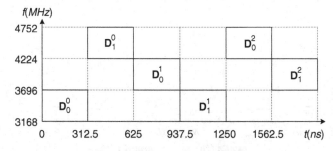

Figure 5.3 Time–frequency representation of multiband OFDM symbols with $K = 2$ and a fast band-hopping rate.

illustrates a time–frequency representation of the signal transmitted, which is based on a time–frequency code that has been proposed for the IEEE 802.15.3a standard [Bat04]. In this example, STF coding is performed across $K = 2$ consecutive OFDM blocks, and the superscript τ of $\mathbf{D}_k^\tau$ represents the index of STF codewords. Since N_b information bits are transmitted in KT_{SYM} seconds, the transmission rate (without channel coding) is $R = N_b/KT_{\text{SYM}}$.

5.2.2 Channel Model

We consider a tap-delay-line Nakagami-m fading channel model with L taps. At the kth OFDM block, the channel impulse response from the ith transmitting antenna to the jth receiver antenna can be described as

$$h_{ij}^k(t) = \sum_{l=0}^{L-1} \alpha_{ij}^k(l)\delta(t - \tau_l), \qquad (5.12)$$

where $\alpha_{ij}^k(l)$ is the multipath gain coefficient, L denotes the number of resolvable paths, and τ_l represents the path delay of the lth path. The amplitude of the lth path, $|\alpha_{ij}^k(l)|$, is modeled as a Nakagami-m random variable with PDF given in (2.11) and average power $\Omega_l = E[|\alpha_{ij}^k(l)|^2]$. The powers of the L paths are normalized such that $\sum_{l=0}^{L-1} \Omega_l = 1$. We assume that the time delay τ_l and the average power Ω_l are the same for every transmitter–receiver link. From (5.12), the channel frequency response is given by

$$H_{ij}^k(f) = \sum_{l=0}^{L-1} \alpha_{ij}^k(l)\exp(-\mathbf{j}2\pi f\tau_l). \qquad (5.13)$$

5.2.3 Receiver Processing

The signal received at each receiver antenna is a superposition of the N_t signals transmitted corrupted by additive white Gaussian noise. Assume that the receiver synchronizes perfectly to the band switching pattern. The RF signal received at each receiver antenna is down-converted to a complex baseband signal, matched to the pulse waveform, and then sampled before passing through an OFDM demodulator. After the OFDM modulator discards the cyclic prefix and performs an N-point FFT, a maximum-likelihood detection is performed jointly across all N_r receiver antennas. The choice of prefix length greater than the duration of the channel impulse response (i.e., $T_{\text{CP}} \geq \tau_{L-1}$) ensures that the interference between OFDM symbols is eliminated. Effectively, the frequency-selective fading channel decouples into a set of N parallel frequency-nonselective fading channels whose fading coefficients are equal to the channel frequency response at the center frequency of the subcarriers. Therefore, the signal received at the nth subcarrier at receiver antenna j during the kth OFDM symbol duration can be expressed as

$$y_j^k(n) = \sqrt{\frac{E}{N_t}} \sum_{i=1}^{N_t} d_i^k(n) H_{ij}^k(n) + z_j^k(n), \tag{5.14}$$

where

$$H_{ij}^k(n) = \sum_{l=0}^{L-1} \alpha_{ij}^k(l) \exp(-\mathbf{j}2\pi n \Delta f \tau_l) \tag{5.15}$$

is the frequency response of the channel at subcarrier n between the ith transmitter and the jth receiver antenna during the kth OFDM block. In (5.14), $z_j^k(n)$ represents the noise sample at the nth subcarrier. We model $z_j^k(n)$ as a complex Gaussian random variable with zero mean and a two-sided power spectral density of $N_0/2$.

For subsequent performance evaluation, we provide a matrix representation of (5.14) as follows. Based on the formulation in [Su05c], we rewrite the signal received at the receiver antenna j in matrix form as

$$\mathbf{Y}_j = \sqrt{\frac{E}{N_t}} \mathbf{S}_D \mathbf{H}_j + \mathbf{Z}_j, \tag{5.16}$$

where $\mathbf{S}_D$ is a $KN \times KNN_t$ data matrix of the form

$$\mathbf{S}_D = [\mathbf{S}_1 \mathbf{S}_2 \cdots \mathbf{S}_{N_t}], \tag{5.17}$$

in which $\mathbf{S}_i$ is a $KN \times KN$ diagonal matrix whose main diagonal comprises the information to be sent from transmitter antenna i. We format $\mathbf{S}_i$ as

$$\mathbf{S}_i = \mathrm{diag}([(\mathbf{d}_i^0)^{\mathrm{T}}(\mathbf{d}_i^1)^{\mathrm{T}} \cdots (\mathbf{d}_i^{K-1})^{\mathrm{T}}]^{\mathrm{T}}), \tag{5.18}$$

where $\mathrm{diag}(\mathbf{x})$ is a diagonal matrix with the elements of $\mathbf{x}$ on its main diagonal. The $KNN_t \times 1$ channel vector $\mathbf{H}_j$ is of the form

$$\mathbf{H}_j = [\mathbf{H}_{1j}^{\mathrm{T}} \mathbf{H}_{2j}^{\mathrm{T}} \cdots \mathbf{H}_{N_t j}^{\mathrm{T}}]^{\mathrm{T}}, \tag{5.19}$$

where

$$\mathbf{H}_{ij} = [H_{ij}^0(0) \cdots H_{ij}^0(N-1) \cdots H_{ij}^{K-1}(0) \cdots H_{ij}^{K-1}(N-1)]^{\mathrm{T}}. \tag{5.20}$$

The received signal $\mathbf{Y}_j$ of size $KNN_r \times 1$ is given by

$$\mathbf{Y}_j = [(\mathbf{y}_j^0)^{\mathrm{T}}(\mathbf{y}_j^1)^{\mathrm{T}} \cdots (\mathbf{y}_j^{K-1})^{\mathrm{T}}]^{\mathrm{T}}, \tag{5.21}$$

in which $\mathbf{y}_j^k$ is an $N \times 1$ vector whose nth element is $y_j^k(n)$. The noise vector $\mathbf{Z}$ has the same form as $\mathbf{Y}$ by replacing $y_j^k(n)$ with $z_j^k(n)$.

We assume that the receiver has perfect knowledge of channel-state information, whereas the transmitter has no channel information. The receiver exploits a maximum likelihood decoder, where the decoding process is jointly performed on N_r receiver signal vectors. The decision rule can be stated as

$$\hat{\mathbf{D}} = \underset{\mathbf{D}}{\arg\min} \sum_{j=1}^{N_r} \left\| \mathbf{Y}_j - \sqrt{\frac{E}{N_t}} \mathbf{S}_D \mathbf{H}_j \right\|^2. \tag{5.22}$$

5.3 PERFORMANCE ANALYSIS

In this section we first present a general framework to analyze the performance of multiband MIMO coding for UWB communication systems. Then we derive the average PEP under the Nakagami-m frequency-selective fading channel model. Finally, we quantify the performance criteria in terms of diversity order and coding gain and determine the maximum achievable diversity advantage for such systems.

Suppose that $\mathbf{D}$ and $\hat{\mathbf{D}}$ are two distinct STF codewords. The PEP, denoted by P_e, is defined as the probability of erroneously decoding the STF codeword $\hat{\mathbf{D}}$ when $\mathbf{D}$ is transmitted. Let $\mathbf{S}_D$ and $\mathbf{S}_{\hat{D}}$ be two data matrices, related to the STF codewords $\mathbf{D}$ and $\hat{\mathbf{D}}$, respectively. Following the computation steps as in [Proa1], the PEP conditioned on the channel matrix is given by

$$P_{e|\mathbf{H}_j} = Q\left(\sqrt{\frac{\rho}{2N_t} \sum_{j=1}^{N_r} \|(\mathbf{S}_D - \mathbf{S}_{\hat{D}})\mathbf{H}_j\|^2} \right), \tag{5.23}$$

where $\rho = E/N_0$ is the average SNR at each receiver antenna and $Q(x)$ is the Gaussian error function as defined in (3.60). The average PEP can be obtained by calculating the expected value of the conditional PEP with respect to the distribution of $\gamma \triangleq \sum_{j=1}^{N_r} \|(\mathbf{S}_D - \mathbf{S}_{\hat{D}})\mathbf{H}_j\|^2$; that is,

$$P_e = \int_0^\infty Q\left(\sqrt{\frac{\rho}{2N_t} s} \right) p_\gamma(s) \, ds, \tag{5.24}$$

where $p_\gamma(s)$ represents the PDF of γ.

For convenience, let us denote an $N_t N_r L K \times 1$ channel vector

$$\mathbf{a} = \left[\mathbf{a}_1^T, \mathbf{a}_2^T \cdots \mathbf{a}_{N_r}^T \right]^T, \tag{5.25}$$

where $\mathbf{a}_j$ contains the multipath gains from all transmitting antennas to the jth receiver antenna. The $N_t L K \times 1$ vector $\mathbf{a}_j$ is formatted as

$$\mathbf{a}_j = \left[\left(\mathbf{a}_{1j}^0 \right)^T \cdots \left(\mathbf{a}_{N_t j}^0 \right)^T \cdots \left(\mathbf{a}_{1j}^{K-1} \right)^T \cdots \left(\mathbf{a}_{N_t j}^{K-1} \right)^T \right]^T, \tag{5.26}$$

in which

$$\mathbf{a}_{ij}^k = \left[\alpha_{ij}^k(0) \, \alpha_{ij}^k(1) \cdots \alpha_{ij}^k(L-1) \right]^T. \tag{5.27}$$

According to (5.15) and (5.26), we can express (5.19) as

$$\mathbf{H}_j = (\mathbf{I}_{KN_t} \otimes \mathbf{W}) \mathbf{a}_j, \tag{5.28}$$

where $\otimes$ denotes the Kronecker product [Hor85], $\mathbf{I}_M$ represents an $M \times M$ identity matrix, and $\mathbf{W}$ is an $N \times L$ Fourier matrix, defined as

$$
\mathbf{W} = \begin{pmatrix}
1 & 1 & \cdots & 1 \\
\omega^{\tau_0} & \omega^{\tau_1} & \cdots & \omega^{\tau_{L-1}} \\
\vdots & \vdots & \ddots & \vdots \\
\omega^{(N-1)\tau_0} & \omega^{(N-1)\tau_1} & \cdots & \omega^{(N-1)\tau_{L-1}}
\end{pmatrix},
\tag{5.29}
$$

in which $\omega = \exp(-\mathbf{j}2\pi\,\Delta f)$. As a consequence, γ can be expressed as

$$
\gamma = \sum_{j=1}^{N_r} \left\| (\mathbf{S}_D - \mathbf{S}_{\hat{D}})(\mathbf{I}_{KN_t} \otimes \mathbf{W})\mathbf{a}_j \right\|^2.
\tag{5.30}
$$

We can see from (5.30) that the distribution of γ depends on the joint distribution of the multipath gain coefficients, $\alpha_{ij}^k(l)$.

In the sequel we evaluate the average PEP of multiantenna multiband OFDM systems, with $|\alpha_{ij}^k(l)|$ being Nakagami-m distributed. First, we analyze the performance of a system with independent fading. Such an assumption allows us to characterize the performances of UWB systems with diversity and coding advantages. The performance of an independent fading system also provides us a benchmark for subsequent performance comparisons. Then we investigate the performance of a more realistic system, where the multipath gain coefficients are allowed to be correlated.

5.3.1 Independent Fading

Due to the band hopping, the K OFDM symbols in each STF codeword are sent over different subbands. With ideal band hopping, we assume that the signal transmitted over K different frequency bands undergo independent fading. We also assume that the path gains $\alpha_{ij}^k(l)$ are independent for different paths and different pairs of transmitting and receiving antennas, and that each transmitting and receiving link has the same power delay profile (i.e., $\mathrm{E}[|\alpha_{ij}^k(l)|^2] = \Omega_l$). The correlation matrix of $\mathbf{a}_j$ is given by

$$
\mathrm{E}\left[\mathbf{a}_j\mathbf{a}_j^{\mathcal{H}}\right] = \mathbf{I}_{KN_t} \otimes \mathbf{\Omega},
\tag{5.31}
$$

where $(\cdot)^{\mathcal{H}}$ denotes a conjugate transpose operation and

$$
\mathbf{\Omega} = \mathrm{diag}(\Omega_0, \Omega_1, \ldots, \Omega_{L-1})
\tag{5.32}
$$

is an $L \times L$ matrix formed from the power of the L paths. Since the matrix $\mathbf{\Omega}$ is diagonal, we can define $\mathbf{\Omega}^{1/2} = \mathrm{diag}(\sqrt{\Omega_0}\sqrt{\Omega_1}\cdots\sqrt{\Omega_{L-1}})$ such that $\mathbf{\Omega} = \mathbf{\Omega}^{1/2}\mathbf{\Omega}^{1/2}$. Let $\mathbf{q}_j = (\mathbf{I}_{KN_t} \otimes \mathbf{\Omega}^{1/2})^{-1}\mathbf{a}^j$; then it is easy to see that the elements of $\mathbf{q}_j$ are independent, identically distributed (iid) Nakagami-m random variables with normalized power $\Omega = 1$. Substitute $\mathbf{a}_j = (\mathbf{I}_{KN_t} \otimes \mathbf{\Omega}^{1/2})\mathbf{q}_j$ into (5.30) and apply the property of the Kronecker product [Hor85, p. 251]:

$$
(\mathbf{A}_1 \otimes \mathbf{B}_1)(\mathbf{A}_2 \otimes \mathbf{B}_2) = (\mathbf{A}_1\mathbf{A}_2) \otimes (\mathbf{B}_1\mathbf{B}_2),
\tag{5.33}
$$

resulting in

$$\gamma = \sum_{j=1}^{N_r} \left\| (\mathbf{S}_D - \mathbf{S}_{\hat{D}})(\mathbf{I}_{KN_t} \otimes \mathbf{W}\mathbf{\Omega}^{1/2})\mathbf{q}_j \right\|^2 = \sum_{j=1}^{N_r} \mathbf{q}_j^{\mathcal{H}} \mathbf{\Psi} \mathbf{q}_j, \qquad (5.34)$$

where

$$\mathbf{\Psi} = (\mathbf{I}_{KN_t} \otimes \mathbf{W}\mathbf{\Omega}^{1/2})^{\mathcal{H}} (\mathbf{S}_D - \mathbf{S}_{\hat{D}})^{\mathcal{H}} (\mathbf{S}_D - \mathbf{S}_{\hat{D}})(\mathbf{I}_{KN_t} \otimes \mathbf{W}\mathbf{\Omega}^{1/2}). \qquad (5.35)$$

Since $\mathbf{\Psi}$ is a Hermitian matrix of size $KN_tL \times KN_tL$, it can be decomposed into $\mathbf{\Psi} = \mathbf{V}\mathbf{\Lambda}\mathbf{V}^{\mathcal{H}}$, where $\mathbf{V} \triangleq \begin{bmatrix} \mathbf{v}_1 \cdots \mathbf{v}_{KN_tL} \end{bmatrix}$ is a unitary matrix, and

$$\mathbf{\Lambda} = \mathrm{diag}\{\lambda_1(\mathbf{\Psi}), \dots, \lambda_{KN_tL}(\mathbf{\Psi})\} \qquad (5.36)$$

is a diagonal matrix whose diagonal elements are the eigenvalues of $\mathbf{\Psi}$. After some manipulations, we arrive at

$$\gamma = \sum_{j=1}^{N_r} \sum_{n=1}^{KN_tL} \lambda_n(\mathbf{\Psi}) |\beta_{j,n}|^2, \qquad (5.37)$$

where $\beta_{j,n} \triangleq \mathbf{v}_n^{\mathcal{H}} \mathbf{q}_j$. Since $\mathbf{V}$ is unitary and the components of $\mathbf{q}_j$ are iid, $\{\beta_{j,n}\}$ are independent random variables whose magnitudes are approximately Nakagami-$\tilde{m}$ distributed with parameter [Nak60, p. 25]

$$\tilde{m} = \frac{KLN_tm}{KLN_tm - m + 1} \qquad (5.38)$$

and average power $\Omega = 1$. Hence, the PDF of $|\beta_{j,n}|^2$ approximately follows a gamma distribution [Sim04, p. 24]

$$p_{|\beta_{j,n}|^2}(x) = \frac{1}{\Gamma(\tilde{m})} \left(\frac{\tilde{m}}{\Omega} \right)^{\tilde{m}} x^{\tilde{m}-1} \exp\left(-\frac{\tilde{m}}{\Omega} x \right). \qquad (5.39)$$

Now the average PEP can be obtained by substituting (5.37) into (5.23) and averaging (5.23) with respect to the distribution of $|\beta_{j,n}|^2$. To this end we resort to an alternative representation of a Q function [Sim04], $Q(x) = (1/\pi) \int_0^{\pi/2} \exp(-x^2/2 \sin^2 \theta) \, d\theta$ for $x \geq 0$. This allows us to express (5.23) in terms of the moment generating function (MGF) of γ, denoted by $\phi_\gamma(s)$, as

$$P_e = \frac{1}{\pi} \int_0^{\pi/2} \phi_\gamma \left(-\frac{\rho}{4N_t \sin^2 \theta} \right) d\theta. \qquad (5.40)$$

Due to the fact that $\phi_{|\beta_{j,n}|^2}(s) = [1 - \Omega/\tilde{m}s]^{-\tilde{m}}$ and $|\beta_{j,n}|^2$ are independent, (5.40) can be written as

$$P_e = \frac{1}{\pi} \int_0^{\pi/2} \prod_{n=1}^{KLN_t} \left(1 + \frac{\rho}{4N_t \sin^2 \theta} \frac{\Omega}{\tilde{m}} \lambda_n(\mathbf{\Psi}) \right)^{-\tilde{m}N_r} d\theta. \qquad (5.41)$$

At high SNR, the average PEP in (5.41) can be upper bounded by

$$P_e \leq \prod_{n=1}^{\text{rank}(\mathbf{\Psi})} \left(\frac{\rho}{4N_t} \frac{\Omega}{\tilde{m}} \lambda_n(\mathbf{\Psi}) \right)^{-\tilde{m}N_r}, \tag{5.42}$$

where $\text{rank}(\mathbf{\Psi})$ and $\{\lambda_n(\mathbf{\Psi})\}_{n=1}^{\text{rank}(\mathbf{\Psi})}$ are the rank and nonzero eigenvalues of matrix $\mathbf{\Psi}$, respectively. In this case, the exponent $\tilde{m}N_r \, \text{rank}(\mathbf{\Psi})$ determines the slope of the performance curve plotted as a function of SNR, whereas the product $(\Omega/\tilde{m})(\prod_{n=1}^{\text{rank}(\mathbf{\Psi})} \lambda_n(\mathbf{\Psi}))^{1/\text{rank}(\mathbf{\Psi})}$ displaces the curve. Therefore, the performance merits of an STF-coded multiband OFDM system can be quantified by the minimum values of these two quantities over all pairs of distinct codewords as the diversity gain

$$G_d = \min_{\mathbf{D} \neq \hat{\mathbf{D}}} \tilde{m} N_r \text{rank}(\mathbf{\Psi}) \tag{5.43}$$

and the coding gain

$$G_c = \min_{\mathbf{D} \neq \hat{\mathbf{D}}} \frac{\Omega}{\tilde{m}} \left(\prod_{n=1}^{\text{rank}(\mathbf{\Psi})} \lambda_n(\mathbf{\Psi}) \right)^{1/\text{rank}(\mathbf{\Psi})}. \tag{5.44}$$

We note that (5.42) can also be derived from the Chernoff bound of the Q function.

To quantify the maximum achievable diversity gain, we calculate the rank of $\mathbf{\Psi}$ as follows. According to (5.35) and the rank property, we have

$$\text{rank}(\mathbf{\Psi}) = \text{rank}((\mathbf{S}_D - \mathbf{S}_{\hat{D}})(\mathbf{I}_{KN_t} \otimes \mathbf{W}\Omega^{1/2})). \tag{5.45}$$

Observe that the size of $\mathbf{S}_D - \mathbf{S}_{\hat{D}}$ is $KN \times KNN_t$, whereas the size of $\mathbf{W}\Omega^{1/2}$ is $N \times L$. Therefore, the rank of matrix $\mathbf{\Psi}$ becomes

$$\text{rank}(\mathbf{\Psi}) \leq \min\{KN, KLN_t\}. \tag{5.46}$$

Hence, the maximum achievable diversity gain is

$$G_d^{\max} = \min\{\tilde{m}KLN_tN_r, \tilde{m}KNN_r\}. \tag{5.47}$$

Note that the diversity gain in (5.47) depends on the parameter $\tilde{m}$, which is close to 1 for any fading parameter m. Indeed, for MIMO multiband OFDM systems,

$$\tilde{m} = \left(1 - \frac{1}{KLN_t} + \frac{1}{KLN_tm} \right)^{-1} \approx 1. \tag{5.48}$$

For example:

- With $N_t = 2$, $K = 2$, $L = 10$, $m = 2$; $\tilde{m} = 1.01 \approx 1$.
- With $N_t = 2$, $K = 2$, $L = 10$, $m = 10$; $\tilde{m} = 1.02 \approx 1$.
- With $N_t = 2$, $K = 2$, $L = 20$, $m = 10$; $\tilde{m} = 1.01 \approx 1$.

In this case, the maximum achievable diversity gain is well approximated by

$$G_d^{\max} = \min\{KLN_tN_r, KNN_r\}. \tag{5.49}$$

The result in the analysis above is somewhat surprising since the diversity gain of a MIMO multiband OFDM system does not depend on the fading parameter m. The reason behind this is that $\beta_{j,n}$ in (5.37) is a normalized summation of KLN_t independent Nakagami random variables. When KLN_t is large enough, $\beta_{j,n}$ behaves like a complex Gaussian random variable, and hence the channel is like Rayleigh fading. Since the ultrawide bandwidth results in a large number of multipath components, the effect of KLN_t on the diversity gain dominates the effect of fading parameter m, and $\tilde{m}$ is close to 1 for any m. This implies that the diversity advantage does not depend on the severity of the fading. The diversity gain obtained under Nakagami fading with an arbitrary m parameter is almost the same as that obtained in Rayleigh fading channels.

We emphasize here the major difference between the use of STF coding in conventional OFDM systems and in the multiband OFDM systems. For STF coding in conventional OFDM systems, the symbols are transmitted continuously in the same frequency band. In this case the temporal diversity relies on the temporal correlation of the channel, and hence the system performance depends on the time-varying nature of the propagation channel [Su05c]. In contrast, the diversity advantage in (5.49) reveals that by the use of band switching, the STF-coded multiband OFDM is able to achieve the diversity gain of $\min\{KLN_tN_r, KNN_r\}$, regardless of the channel time-correlation property.

It is worth noting that the theoretical framework presented incorporates the analysis for ST- or SF-coded UWB systems as special cases. For a single-carrier frequency-nonselective channel (i.e., $N = 1$ and $L = 1$), the performance of an STF-coded UWB system is similar to that of an ST-coded UWB system. When $K = 1$ (i.e., when the coding is performed over one OFDM block), the STF-coded UWB system performance is the same as that of an SF-coded scheme. The maximum achievable diversity reduces to $\min\{LN_tN_r, NN_r\}$. This reveals that as long as the K OFDM symbols are sent on different frequency bands, coding across K OFDM blocks can offer a diversity advantage K times larger than that from an SF coding approach.

5.3.2 Correlated Fading

In this section we investigate the performance of STF-coded multiband OFDM systems in correlated fading scenarios. From (5.30) we know that γ can be expressed as

$$\gamma = \mathbf{a}^{\mathcal{H}}\{\mathbf{I}_{N_r} \otimes [(\mathbf{I}_{KN_t} \otimes \mathbf{W}^{\mathcal{H}})(\mathbf{S}_D - \mathbf{S}_{\hat{D}})^{\mathcal{H}}(\mathbf{S}_D - \mathbf{S}_{\hat{D}})(\mathbf{I}_{KN_t} \otimes \mathbf{W})]\}\mathbf{a}. \qquad (5.50)$$

To simplify the analysis, we assume that the channel correlation matrix $\mathbf{R}_A = \mathrm{E}[\mathbf{a}\mathbf{a}^{\mathcal{H}}]$ is of full rank. Since $\mathbf{R}_A$ is positive-definite Hermitian symmetric, it has a symmetric square root $\mathbf{U}$ such that $\mathbf{R} = \mathbf{U}^{\mathcal{H}}\mathbf{U}$, where $\mathbf{U}$ is also of full rank [Hor85]. Let $\mathbf{q} = \mathbf{U}^{-1}\mathbf{a}$; it follows that $\mathrm{E}[\mathbf{q}\mathbf{q}^{\mathcal{H}}] = \mathbf{I}_{KLN_tN_r}$ (i.e., the components of $\mathbf{q}$ are uncorrelated). Substituting $\mathbf{a} = \mathbf{U}\mathbf{q}$ into (5.50), we have

$$\gamma = \mathbf{q}^{\mathcal{H}}\boldsymbol{\Phi}\mathbf{q}, \qquad (5.51)$$

where

$$\Phi = U^{\mathcal{H}}\{I_{N_r} \otimes [(I_{KN_t} \otimes W^{\mathcal{H}})(S_D - S_{\hat{D}})^{\mathcal{H}}(S_D - S_{\hat{D}})(I_{KN_t} \otimes W)]\}U. \qquad (5.52)$$

Accordingly, performing an eigenvalue decomposition of the $KLN_tN_r \times KLN_tN_r$ Hermitian symmetric matrix Φ results in $\Phi = V\Lambda V^{\mathcal{H}}$. Therefore, we can express (5.51) as

$$\gamma = \sum_{n=1}^{KLN_tN_r} \lambda_n(\Phi)|\beta_n|^2, \qquad (5.53)$$

where $\beta_n \triangleq v_n^{\mathcal{H}}q$ and the v_n's and $\lambda_n(\Phi)$'s are eigenvectors and eigenvalues of matrix Φ. From (5.24) and (5.53), the PEP can be obtained by averaging the conditional PEP with respect to the joint distribution of $\{|\beta_n|^2\}$; that is,

$$P_e = \int_0^{\infty} \cdots \int_0^{\infty} Q\left(\sqrt{\frac{\rho}{2N_t}\sum_{n=1}^{M} \lambda_n(\Phi)s_n}\right) p_{|\beta_1|^2\cdots|\beta_M|^2}(s_1, \ldots, s_M)ds_1 \cdots ds_M,$$

$$(5.54)$$

where $M = KLN_tN_r$. In general, β_n's for different n are not independent, and the closed-form solution for (5.54) is difficult, if not possible, to determine. In what follows we discuss two special cases where the average PEP in (5.54) can be simplified further.

Special Case 1: Constant Fading We consider the situation when the MIMO channel stays constant over K OFDM blocks. This corresponds to the case when the modulated OFDM signal is transmitted continually over the same subband for entire K OFDM symbol periods. Figure 5.4 illustrates such a multiband signal with one of the time–frequency codes in the IEEE 802.15.3a standard proposal [Bat04]. In this example, STF coding is applied across $K = 2$ OFDM blocks and two OFDM symbols are sent on one subband before the band switching.

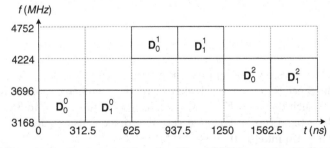

Figure 5.4 Time–frequency representation of multiband OFDM symbols with $K = 2$ and a slow band-hopping rate.

In this case, (5.30) can be reexpressed as

$$\gamma = \sum_{j=1}^{N_r} \left\| (\mathbf{C}_D - \mathbf{C}_{\hat{D}})(\mathbf{I}_{N_t} \otimes \mathbf{W})\tilde{\mathbf{a}}_j \right\|^2, \tag{5.55}$$

where $\mathbf{C}_D = [\mathbf{C}^T{}_0 \mathbf{C}^T{}_1 \cdots \mathbf{C}^T{}_{K-1}]^T$ is a $KN \times N_t N$ matrix and

$$\mathbf{C}_k = \left[\mathrm{diag}(\mathbf{d}_1^k)\mathrm{diag}(\mathbf{d}_2^k) \cdots \mathrm{diag}(\mathbf{d}_{N_t}^k) \right]. \tag{5.56}$$

The channel vector $\tilde{\mathbf{a}}_j$ of size $LN_t \times 1$ is given by $\tilde{\mathbf{a}}_j = [\mathbf{a}_{1j}^T \mathbf{a}_{2j}^T \cdots \mathbf{a}_{N_t j}^T]^T$, in which $\mathbf{a}_{ij}$ is defined in (5.27). Since the path gains $\mathbf{a}_{ij}^k$'s are the same for every k, $0 \le k \le K-1$, the time superscript index k is omitted to simplify the notations. Following the steps given previously, we can show that the average PEP is of a form similar to (5.54), with M replaced by $LN_t N_r$ and $\{\lambda_n(\tilde{\mathbf{\Phi}})\}_{n=1}^{LN_t N_r}$ being eigenvalues of the matrix

$$\tilde{\mathbf{\Phi}} = \tilde{\mathbf{U}}^{\mathcal{H}} \{ \mathbf{I}_{N_r} \otimes [(\mathbf{I}_{N_t} \otimes \mathbf{W}^{\mathcal{H}})(\mathbf{C}_D - \mathbf{C}_{\hat{D}})^{\mathcal{H}}(\mathbf{C}_D - \mathbf{C}_{\hat{D}})(\mathbf{I}_{N_t} \otimes \mathbf{W})] \} \tilde{\mathbf{U}}. \tag{5.57}$$

Here $\tilde{\mathbf{U}}$ is a symmetric square root of $\tilde{\mathbf{R}}_A = \mathrm{E}[\tilde{\mathbf{a}}\tilde{\mathbf{a}}^{\mathcal{H}}]$, in which $\tilde{\mathbf{a}} = [\tilde{\mathbf{a}}_1^T \tilde{\mathbf{a}}_2^T \cdots \tilde{\mathbf{a}}_{N_r}^T]^T$.

With a further assumption that the path gains are independent for every transmitter–receiver link, the average PEP can be obtained in a fashion similar to that derived in Section 5.31 as

$$P_e \le \left[\prod_{n=1}^{\mathrm{rank}(\Theta)} \left(\frac{\rho}{4N_t} \frac{\Omega}{\tilde{m}} \lambda_n(\Theta) \right) \right]^{-\tilde{m} N_r}, \tag{5.58}$$

where $\lambda_n(\Theta)$'s are the nonzero eigenvalues of the matrix

$$\Theta = (\mathbf{I}_{N_t} \otimes \mathbf{W}^{\mathcal{H}})(\mathbf{C}_D - \mathbf{C}_{\hat{D}})^{\mathcal{H}}(\mathbf{C}_D - \mathbf{C}_{\hat{D}})(\mathbf{I}_{N_t} \otimes \mathbf{W}). \tag{5.59}$$

Observe that the maximum rank of $(\mathbf{C}_D - \mathbf{C}_{\hat{D}})(\mathbf{I}_{N_t} \otimes \mathbf{W})$ is $\min\{LN_t, KN\}$. In typical multiband OFDM systems, the number of subcarriers, N, is larger than LN_t; hence, the maximum achievable diversity gain of this system is $LN_t N_r$. Based on this observation, we can conclude that when K OFDM symbols are sent on one subband prior to band switching, coding across K OFDM blocks does not offer any additional diversity advantage compared to the coding scheme within one OFDM block.

Special Case 2: Fading Parameter $m = 1$ With $m = 1$, Nakagami is equivalent to Rayleigh distribution, and the path gain coefficients can be modeled as complex Gaussian random variables. Recall that for Gaussian random variables, uncorrelated implies independent. Thus, $\{|\beta_n|^2\}$ in (5.53) becomes a set of iid Rayleigh random variables. By using an MGF of γ, the average PEP in (5.54) is given by

$$P_e = \frac{1}{\pi} \int_0^{\pi/2} \prod_{n=1}^{KLN_t N_r} \left(1 + \frac{\rho}{4N_t \sin^2 \theta} \lambda_n(\mathbf{\Phi}) \right)^{-1} d\theta. \tag{5.60}$$

where Φ is defined in (5.51). The PEP above can be upper-bounded by

$$P_e \leq \left[\prod_{n=1}^{KLN_tN_r} \left(\frac{\rho}{4N_t} \lambda_n(\Phi) \right) \right]^{-1} \tag{5.61}$$

at high SNR. Therefore, the diversity and coding gains for this system are defined, respectively, as

$$G_d = \min_{\mathbf{D} \neq \hat{\mathbf{D}}} N_r \ \text{rank}(\Phi) \tag{5.62}$$

and

$$G_c = \min_{\mathbf{D} \neq \hat{\mathbf{D}}} \left(\prod_{n=1}^{\text{rank}(\Phi)} \lambda_n(\Phi) \right)^{1/\text{rank}(\Phi)}. \tag{5.63}$$

5.4 SIMULATION RESULTS

To support the theoretical analysis given in preceding sections, we perform simulations for multiantenna multiband OFDM systems employing various STF codes. Following the IEEE 802.15.3a standard proposal, our simulated multiband OFDM system has $N = 128$ subcarriers, and the bandwidth of each subband is BW = 528 MHz. Thus, the OFDM symbol is of duration $T_{\text{FFT}} = 128/(528 \text{ MHz}) = 242.42$ ns. After adding the cyclic prefix of length $T_{\text{CP}} = 60.61$ ns and the guard interval of length $T_{\text{GI}} = 9.47$ ns, the symbol duration becomes $T_{\text{SYM}} = 312.5$ ns.

We simulated the STF-coded multiband OFDM systems in a Nakagami-m fading environment. We employed the stochastic tapped-delay-line channel model in (5.12), where the path amplitudes $|\alpha_{ij}^k(l)|$ are Nakagami-m distributed and the phases $\angle\alpha_{ij}^k(l)$ are chosen uniformly from $[0, 2\pi)$. The path gain coefficients $\alpha_{ij}^k(l)$ for different i, j, and l are generated independently. The power delay profile, used to specify the path delays τ_l's and powers Ω_l's, follows the statistical model in [Tar03], which is based on an extensive propagation study in residential environments. Figure 5.5 shows the power delay profile of the simulated channel. Note that in our simulations, we normalize the total average power of the L paths to unity (i.e., $\sum_{l=0l}^{L-1} \Omega_l = 1$).

In our simulations, the STF codeword $\mathbf{D} = [\mathbf{D}_0^T \mathbf{D}_1^T \cdots \mathbf{D}_{K-1}^T]^T$ in (5.8) is further simplified as

$$\mathbf{D}_k = \left[\mathbf{G}_{k,1}^T \ \mathbf{G}_{k,2}^T \ \cdots \ \mathbf{G}_{k,P}^T \ \mathbf{0}_{(N-P\Upsilon N_t) \times N_t}^T \right], \tag{5.64}$$

in which Υ is a fixed integer ($\Upsilon \in \{1, 2, \ldots, L\}$), $P = \lfloor N/\Upsilon N_t \rfloor$, and $\mathbf{0}_{m \times n}$ stands for an $m \times n$ all-zero matrix. The code matrix $\mathbf{G}_{k,p}$ for $p = 1, 2, \ldots, P$ and $k = 0, 1, \ldots, K - 1$ is of size $\Upsilon N_t \times N_t$. The code matrices $\{\mathbf{G}_{k,p}\}_{k=0}^{K-1}$ for each p are designed jointly, whereas the matrices $\mathbf{G}_{k,p}$ and $\mathbf{G}_{k',p'}$ with $p \neq p'$ are designed independently. Such code structures are able to provide the maximum achievable diversity while enabling low computational complexity [Su05c].

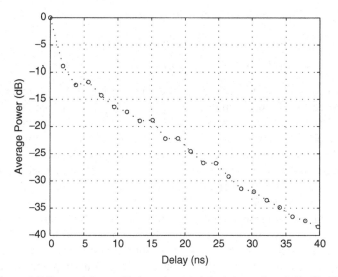

Figure 5.5 Power delay profile based on statistical channel model in [Tar03].

Let us consider a system with two transmitting antennas. Based on the repetition STF code in [Su05c], $\mathbf{G}_{k,p}$ is given by

$$\mathbf{G}_{k,p} = \left(\mathbf{I}_{N_t} \otimes \mathbf{1}_{\Upsilon \times 1}\right) \begin{pmatrix} x_{p,1} & x_{p,2} \\ -x_{p,2}^* & x_{p,1}^* \end{pmatrix}, \tag{5.65}$$

where $\mathbf{1}_{m \times n}$ denotes an $m \times n$ all-1 matrix, and $x_{p,i}$'s are selected from BPSK or QPSK constellations. Note that $\mathbf{G}_{k,p}$ is the same for all k's. We also exploit a full-rate STF code [Su05c] in which $\mathbf{G}_{k,p}$ is

$$\mathbf{G}_{k,p} = \sqrt{N_t} \begin{pmatrix} \mathbf{x}_{p,1}^k & \mathbf{0}_{\Upsilon \times 1} \\ \mathbf{0}_{\Upsilon \times 1} & \mathbf{x}_{p,2}^k \end{pmatrix}. \tag{5.66}$$

In (5.66), $\mathbf{x}_{p,i}^k$ is a column vector of length Υ whose elements are specified as follows. For notation convenience, we omit the subscript p and denote $\mathcal{L} = K \Upsilon N_t$. Let $\mathbf{s} = [s_1 s_2 \cdots s_{\mathcal{L}}]$ be a vector of BPSK or QPSK symbols. The $1 \times \mathcal{L}$ matrix

$$\mathbf{x} \triangleq \left[\left(\mathbf{x}_1^0\right)^\mathsf{T} \left(\mathbf{x}_2^0\right)^\mathsf{T} \cdots \left(\mathbf{x}_1^{K-1}\right)^\mathsf{T} \left(\mathbf{x}_2^{K-1}\right)^\mathsf{T} \right] \tag{5.67}$$

is given by

$$\mathbf{x} = \frac{1}{\sqrt{K}} \mathbf{s} \frac{1}{\sqrt{K}} \mathbf{V}(\theta_1, \theta_2, \ldots, \theta_{\mathcal{L}}), \tag{5.68}$$

in which **V** is a Vandermonde matrix[1] with [Su05c]

$$\theta_l = \begin{cases} \exp(j(4l-3)\pi/(2\mathcal{L})) & \text{for } \mathcal{L} = 2^s (s \geq 1); \\ \exp(j(6l-1)\pi/(3\mathcal{L})) & \text{for } \mathcal{L} = 2^s \cdot 3^t (s \geq 0, t \geq 1). \end{cases} \tag{5.69}$$

We note that when $K = 1$, the repetition-coded and full-rate STF codes are reduced to those proposed for SF coding in [Su03, Su05a, Su04a]. Unless specified otherwise, we apply a random permutation technique [Su04a] so as to reduce the correlation in the channel frequency response among different subcarriers. This permutation strategy allows us to achieve a larger coding advantage and hence improve system performance. Note that our simulation results are based on uncoded information. The performance can be further improved by the use of channel coding, such as convolutional and Reed–Solomon codes [Bat04].

In what follows we present the average bit error rate (BER) curves of multiband OFDM systems as functions of the average SNR per bit (E_b/N_0) in decibels. In every case, curves with circles, crosses, and triangles show the performance of systems with single transmitter and single receiver antennas, two transmitting and one receiving antennas, and two transmitting and two receiving antennas, respectively.

First, we consider the performance of a coding approach over one OFDM block ($K = 1$). We utilize both repetition-coded and full-rate STF codes, each with a spectral efficiency of 1 bit/s·Hz (omitting the prefix and guard interval). We use QPSK constellation for the repetition code and BPSK for the full-rate STF code. Both systems achieve a data rate (without channel coding) of 128 bits/312.5 ns = 409.6 Mbps. Figure 5.6 depicts the performances of the STF-coded UWB system with $\Upsilon = 2$. We observe that regardless of a particular STF coding scheme, the spatial diversity gained from multiantenna architecture does improve system performance significantly. In addition, the performance can be improved further with the choice of STF codes and permutation schemes. In Fig. 5.7 we compare the performance of a multiband OFDM system with different frequency-diversity orders. Here we employ a full-rate code with $\Upsilon = 2, 3$, and 4. We can see that by increasing the number of jointly encoded subcarriers, system performance can be improved. This observation is in accordance with our theoretical result in (5.42). Therefore, with a properly designed STF code, we can effectively exploit both spatial and frequency diversities in a UWB environment.

Second, we compare the performances of STF-coded multiband OFDM system, in which the coding is performed over one and two OFDM blocks ($K = 1, 2$). We consider a scenario when two consecutive OFDM symbols are transmitted over different subbands- for instance, when the multiband signal has a time–frequency representation as in Fig. 5.3. The performances of the repetition and full-rate STF-coded UWB systems with $\Upsilon = 2$ are shown in Fig. 5.8(a) and (b), respectively. The repetition code is constructed from BPSK constellation for $K = 1$ and QPSK for $K = 2$. Thus, the spectral efficiency of the resulting codes is 0.5 bit/s·Hz. Full-rate STF

---

[1] A Vandermonde matrix with variables $\theta_1, \theta_2, \ldots, \theta_{\mathcal{L}}$ is a $\mathcal{L} \times \mathcal{L}$ matrix whose lth ($l = 1, 2, \ldots, \mathcal{L}$) row is defined as $[\theta_1^{l-1} \theta_2^{l-1} \cdots \theta_{\mathcal{L}}^{l-1}]$.

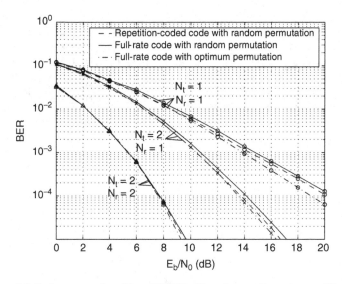

Figure 5.6 Performance of multiband OFDM with various coding schemes ($K = 1$).

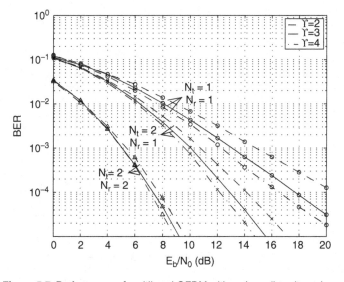

Figure 5.7 Performance of multiband OFDM with various diversity orders.

codes are generated from BPSK constellation for both $K = 1$ and 2, and their spectral efficiency is 1 bit/s·Hz. From both figures it is apparent that by jointly coding over multiple OFDM blocks, an STF-coded UWB system has a BER performance curve that is steeper than that of a UWB system without joint encoding (i.e., the diversity advantage increases with the number of jointly encoded OFDM blocks). Such

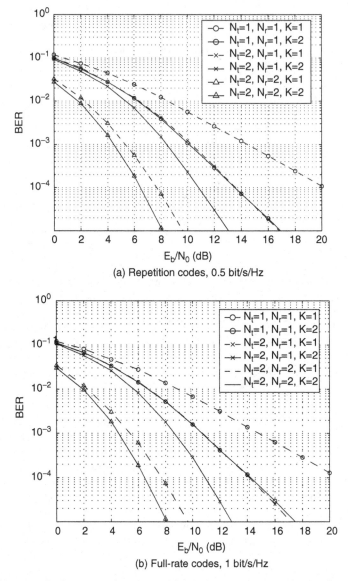

Figure 5.8 Performance of multiband OFDM with various time spreading factors.

improvement results from band hopping rather than temporal diversity. Hence, by coding across multiple OFDM blocks, the diversity order of STF-coded band-hopping UWB increases significantly regardless of the temporal correlation of the channel. This supports our analytical results in Section 5.3 that the diversity order of an STF-coded multiband OFDM system with a fast band-hopping rate is increasing with K.

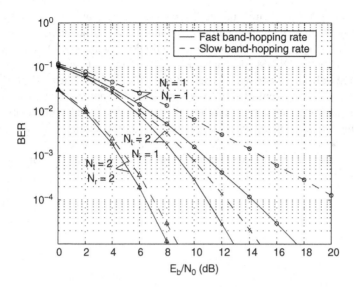

Figure 5.9 Performance of multiband OFDM with various hopping rates.

Finally, we compare the performance of multiband systems with different band-hopping rates. Figure 5.9 depicts the performance of a full-rate STF-coded UWB system with $\Upsilon = 2$ and $K = 2$. Each STF codeword is transmitted during two OFDM block periods. We consider the cases when two consecutive OFDM symbols are sent on a different subband (fast band-hopping rate), and when they are transmitted continually on the same frequency band (slow band-hopping rate). From Fig. 5.9 we observe the performance degradation when the band-hopping rate decreases, which corresponds to the results in (5.42) and (5.58) that coding over multiple OFDM blocks will offer an additional diversity advantage when the STF coding is applied together with a fast band-hopping scheme (i.e., the K OFDM symbols in each STF codeword are transmitted on various frequency bands).

5.5 CHAPTER SUMMARY

In conventional OFDM systems with N_t transmitting and N_r receiving antennas, STF coding across K OFDM blocks can lead to a maximum achievable diversity order of TLN_tN_r, where L is the number of resolvable paths and T is the rank of the temporal correlation matrix of the channel ($T \leq K$). In this chapter we present a multiband MIMO coding framework for UWB systems. By a technique of band hopping in combination with joint coding across spatial, temporal, and frequency domains, the scheme is able to exploit all available spatial and multipath diversities, richly inherent in UWB environments. From the theoretical results we can draw some interesting conclusions. First, the effect of Nakagami fading parameter m on the diversity gain is insignificant, and the diversity advantages obtained in Nakagami-m and Rayleigh fading channels are almost the same. Second, the maximum achievable diversity advantage of a MIMO multiband OFDM system is KLN_tN_r. In contrast to the conventional OFDM, the factor K comes from the band-hopping approach, which is regardless of the temporal correlation of the channel. The simulation results show that the employment of STF coding and band-hopping techniques is able to increase the diversity advantage significantly, thereby considerably improving system performance. In a single-antenna system, increasing the number of jointly encoded OFDM blocks from one to two yields a performance improvement of 6 dB at a BER of 10^{-4}. By also increasing the number of transmitting antennas from one to two, the STF-coded multiband OFDM system has a total gain of 9 dB at a BER of 10^{-4}.

6

PERFORMANCE
CHARACTERIZATION

In recent years, the performance analysis of UWB systems has been an area of considerable interest. Many papers deal with the performance of UWB systems over additive white Gaussian noise (AWGN) as well as fading channels. For example, the performance of a single-user time-hopping UWB system over multipath channels corrupted by AWGN was analyzed in [Cho02]. The authors in [Dur03b] evaluated the bit error probability of a time-hopping UWB system in an AWGN channel with the presence of multiuser interference. Later in [Bou04], the authors analyzed the average BER performance of a multiuser direct-sequence UWB system over a lognormal fading channel. The authors in [Zha03] derived an explicit symbol error probability expression for a UWB system employing a RAKE receiver in multipath lognormal fading channels. In [Cha04] the BER of a time-hopping system in multipath Rayleigh fading channels was discussed. More recently, the performance of a multiantenna single-band UWB system has been analyzed in multipath Nakagami-m fading channels: The capacity analysis is provided in [Fen04] and the BER analysis is given in [Sir05a]. The error probability performance of a multiband OFDM system under Nakagami-m fading channels is analyzed in Chapter 5.

Although a clustering phenomenon has been observed in several large data sets of UWB channel measurements [TG3a], it has not been taken into consideration for the analysis due to the fact that random clustering behavior introduces a difficulty in evaluating the analytical performance. In fact, most existing work is based on stochastic tapped-delay-line (STDL) models [Pro01] used in conventional narrowband and wideband systems. However, performance analysis in STDL models is basically an extension of that for narrowband systems. More important, it does not reflect the multipath-rich nor random-clustering characteristics of UWB channels. To the best of our knowledge, none of the existing analysis is insightful in revealing the effect of the unique clustering characteristic on UWB system performance. To implement

Ultra-Wideband Communications Systems: Multiband OFDM Approach, By W. Pam Siriwongpairat and K. J. Ray Liu
Copyright © 2008 John Wiley & Sons, Inc.

Adapted with permission from © 2006 *IEEE Journal on Selected Area in Communications*, Vol. 24, no. 4, Part 1, Apr. 2006, pp. 745–751.

an efficient UWB system, it is vital to capture the behavior of UWB channels, which has been characterized by the S-V model [Sal87].

In this chapter we analyze the performance of UWB systems employing multi-band OFDM [Bat04] by taking into account the multipath-rich and random-clustering characteristics of UWB channels. Using the S-V model, we characterize the UWB performance in terms of cluster arrival rate, ray arrival rate, and cluster and ray decay factors [Sir05d, Sir06b]. Two performance criteria we consider are PEP and outage probability. First, we provide an exact PEP formulation for single-antenna multiband OFDM systems. Then we establish an approximation approach, which allows us to obtain a closed-form PEP formulation and an explicit outage probability expression. It turns out that the uncoded multiband OFDM system cannot gain from the multipath-clustering property of UWB channels. On the other hand, jointly encoding the data across subcarriers yields performance improvement, which depends strongly on cluster and ray arrival rates. Finally, we generalize the performance results to STF-coded MIMO multiband OFDM systems. The theoretical analysis reveals that the diversity gain does not rely heavily on the clustering phenomenon of UWB channels, whereas the coding gain is in terms of both multipath arrival rates and decay factors. Simulation results are provided to support the theoretical analysis.

A brief description of the channel model and system model under consideration are given in Section Section 6.1. PEP and outage probability analysis is provided in Section 6.2. First, an approximation technique is established and a new closed-form PEP formulation is obtained. Then we present closed-form expressions for the probability density function (PDF) and outage probability of the combined signal-to-noise ratio over S-V fading scenarios. In Section 6.3 we characterize the performance of MIMO multiband OFDM systems. Simulation results are presented in Section 6.4.

6.1 SYSTEM MODEL

We consider a peer-to-peer UWB multiband OFDM system. Within each subband, the information is modulated using OFDM with N subcarriers. The channel model is based on the S-V model as described in (2.14).

The path amplitude $|\alpha(c, l)|$ may follow the lognormal distribution [Foe03b], the Nakagami distribution [Win02], or the Rayleigh distribution [Cra02], whereas the phase $\angle\alpha(c, l)$ is uniformly distributed over $[0, 2\pi)$. For analytical tractability and to obtain insight into UWB systems, we consider the scenario that the path amplitude $|\alpha(c, l)|$ is modeled as a Rayleigh distribution [Foe03b, Cra02]. Specifically, the multipath gain coefficients $\alpha(c, l)$ are modeled as zero-mean, complex Gaussian random variables with variances $\Omega_{c,l}$ as specified in (2.17). The powers of the multipath components are normalized such that $\sum_{c=0}^{C} \sum_{l=0}^{L} \Omega_{c,l} = 1$. The channel parameters corresponding to different fading scenarios are specified in [Foe03b]. From (2.14), the channel frequency response is given by

$$H(f) = \sum_{c=0}^{C} \sum_{l=0}^{L} \alpha(c, l) \exp[-\mathbf{j}2\pi f(T_c + \tau_{c,l})]. \qquad (6.1)$$

With the choice of cyclic prefix length greater than the duration of the channel impulse response, OFDM allows for each UWB subband to be divided into a set of N orthogonal narrowband channels. At the transmitter, an information sequence is partitioned into blocks. Each block is mapped onto an $N \times 1$ matrix

$$\mathbf{D} = [d(0)\,d(1)\,\cdots\,d(N-1)]^{\mathrm{T}}, \tag{6.2}$$

where $d(n)$, $n = 0, 1, \ldots, N-1$, represents the complex channel symbol to be transmitted over subcarrier n. Suppose that the information is jointly encoded across M ($1 \leq M \leq N$) subcarriers. Particularly, the data matrix $\mathbf{D}$ is a concatenation of $P = \lfloor N/M \rfloor$ data matrices as follows:

$$\mathbf{D} = [\mathbf{D}_0^{\mathrm{T}}\mathbf{D}_1^{\mathrm{T}}\,\cdots\,\mathbf{D}_{(P-1)}^{\mathrm{T}}\mathbf{0}_{(N-PM)\times 1}]^{\mathrm{T}}, \tag{6.3}$$

where $\mathbf{D}_p$ is an $M \times 1$ data matrix defined as

$$\mathbf{D}_p = [d_p(0)d_p(1)\cdots d_p(M-1)]^{\mathrm{T}} \tag{6.4}$$

with

$$d_p(n) \triangleq d(pM+n); \quad p = 0, 1, \ldots, P-1, \tag{6.5}$$

and $\mathbf{0}_{m\times n}$ stands for an $m \times n$ all-zero matrix. The data matrices $\mathbf{D}_p$ are designed independently for different p's. Also, each data symbol $d_p(n)$ is normalized to have unit energy (i.e., the data matrix satisfies the energy constraint $\mathrm{E}[\|\mathbf{D}_p\|^2] = M$ for all p). The transmitter applies an N-point IFFT to the matrix $\mathbf{D}$, appends a cyclic prefix and guard interval, up-converts to RF, and then sends the signal modulated at each subcarrier.

At the receiver, after matched filtering, removing the cyclic prefix, and applying FFT, the signal received at the nth subcarrier is given by

$$y(n) = \sqrt{E_s}d(n)H(n) + z(n), \tag{6.6}$$

where E_s is the average energy transmitted per symbol, $H(n)$ is the channel frequency response at the nth subcarrier, and $z(n)$ denotes the noise sample at the nth subcarrier. We model $z(n)$ as a complex Gaussian random variable with zero mean and variance N_0. The channel frequency response can be specified as

$$H(n) = \sum_{c=0}^{C}\sum_{l=0}^{L}\alpha(c, l)\exp[-\mathbf{j}2\pi n\Delta f(T_c + \tau_{c,l})], \tag{6.7}$$

where $\Delta f = 1/T$ is the frequency separation between two adjacent subcarriers and T is the OFDM symbol period. We assume that the channel-state information $H(n)$ is known at the receiver but not at the transmitter.

6.2 PERFORMANCE ANALYSIS

In this section we present at first a general framework to analyze the PEP performance of UWB multiband OFDM systems. Then, using the S-V model, we characterize the average PEP of UWB systems in terms of multipath arrival rates and decay factors. Finally, an outage probability formulation of UWB systems in S-V fading channel is provided.

6.2.1 Average PEP Analysis

For subsequent performance evaluation, we format the received signal in (6.6) in vector form as

$$\mathbf{Y}_p = \sqrt{E_s} X(\mathbf{D}_p)\mathbf{H}_p + \mathbf{Z}_p, \tag{6.8}$$

where

$$X(\mathbf{D}_p) = \mathrm{diag}(d_p(0), d_p(1), \ldots, d_p(M-1)) \tag{6.9}$$

is an $M \times M$ diagonal matrix with the elements of $\mathbf{D}_p$ on its main diagonal. In (6.8), the channel matrix $\mathbf{H}_p$, the received signal matrix $\mathbf{Y}_p$, and the noise matrix $\mathbf{Z}_p$ are of the same forms as $\mathbf{D}_p$ by replacing d with H, y, and z, respectively. The receiver exploits a maximum likelihood decoder, where the decoding process is performed jointly within each data matrix $\mathbf{D}_p$, and the decision rule can be stated as

$$\hat{\mathbf{D}}_p = \underset{\mathbf{D}_p}{\mathrm{argmin}} \, \| \mathbf{Y}_p - \sqrt{E_s} X(\mathbf{D}_p)\mathbf{H}_p \|^2. \tag{6.10}$$

Suppose that $\mathbf{D}_p$ and $\hat{\mathbf{D}}_p$ are two distinct data matrices. Since the data matrices $\mathbf{D}_p$ for different p's are en/decoded independently, for simplicity the PEP can be defined as the probability of erroneously decoding the matrix $\hat{\mathbf{D}}_p$ when $\mathbf{D}_p$ is transmitted. Following the computation steps as in [Pro01], the average PEP denoted as P_e, is given by

$$P_e = \mathrm{E}\left[Q\left(\sqrt{\frac{\rho}{2} \| \mathbf{\Delta}_p \mathbf{H}_p \|^2} \right) \right], \tag{6.11}$$

where $\rho = E_s/N_0$ is the average SNR,

$$\mathbf{\Delta}_p = X(\mathbf{D}_p) - X(\hat{\mathbf{D}}_p) \tag{6.12}$$

is the signal difference matrix, and $Q(\cdot)$ is the Gaussian error function, as defined in (3.60). According to [Sim00], the Gaussian error function in (3.60) can also be expressed as

$$Q(x) = \frac{1}{\pi} \int_0^{\pi/2} \exp\left(-\frac{x^2}{2\sin^2\theta} \right) d\theta; \quad x \geq 0. \tag{6.13}$$

Denoting

$$\eta = \| \mathbf{\Delta}_p \mathbf{H}_p \|^2, \tag{6.14}$$

and using an alternative representation of the Q function in (6.13), the average PEP in (6.11) can be expressed as

$$P_e = \frac{1}{\pi} \int_0^{\pi/2} \mathcal{M}_\eta \left(-\frac{\rho}{4 \sin^2 \theta} \right) d\theta, \tag{6.15}$$

where

$$\mathcal{M}_\eta(s) = \mathrm{E}\left[\exp(s\eta) \right] \tag{6.16}$$

represents the MGF of η [Sim00]. From (6.15) we can see that the remaining problem is to obtain the MGF $\mathcal{M}_\eta(s)$.

For convenience, let us denote a $(C+1)(L+1) \times 1$ channel matrix

$$\mathbf{A} = [\alpha(0,0)\cdots\alpha(0,L)\cdots\alpha(C,0)\cdots\alpha(C,L)]^{\mathrm{T}}. \tag{6.17}$$

According to (6.7), $\mathbf{H}_p$ can be decomposed as

$$\mathbf{H}_p = \mathbf{W}_p \cdot \mathbf{A}, \tag{6.18}$$

where

$$\mathbf{W}_p = \begin{pmatrix} \omega_{p,0}^{T_0+\tau_{0,0}} & \omega_{p,0}^{T_0+\tau_{0,1}} & \cdots & \omega_{p,0}^{T_C+\tau_{C,L}} \\ \omega_{p,1}^{T_0+\tau_{0,0}} & \omega_{p,1}^{T_0+\tau_{0,1}} & \cdots & \omega_{p,1}^{T_C+\tau_{C,L}} \\ \vdots & \vdots & \ddots & \vdots \\ \omega_{p,M-1}^{T_0+\tau_{0,0}} & \omega_{p,M-1}^{T_0+\tau_{0,1}} & \cdots & \omega_{p,M-1}^{T_C+\tau_{C,L}} \end{pmatrix},$$

in which

$$\omega_{p,n} \triangleq \exp[-\mathrm{j}2\pi \, \Delta f(pM+n)]. \tag{6.19}$$

After some manipulation, we can rewrite η in (6.14) as

$$\eta = \sum_{n=1}^{M} \mathrm{eig}_n(\mathbf{\Psi})|\beta_n|^2, \tag{6.20}$$

where β_n are iid complex Gaussian random variables with zero mean and unit variance, and $\mathrm{eig}_n(\mathbf{\Psi})$ are the eigenvalues of the matrix

$$\mathbf{\Psi} = \|\mathbf{\Delta}_p \mathbf{W}_p \mathbf{\Omega}^{1/2}\|^2, \tag{6.21}$$

in which

$$\mathbf{\Omega} = \mathrm{diag}(\Omega_{0,0}, \Omega_{0,1}, \ldots, \Omega_{C,L}) \tag{6.22}$$

is a diagonal matrix formed from the average powers of multipath components. Thus, the MGF of η in (6.20) can be given by

$$\mathcal{M}_\eta(s) = \mathrm{E}\left[\prod_{n=1}^{M}(1 - s \, \mathrm{eig}_n(\mathbf{\Psi}))^{-1} \right]. \tag{6.23}$$

Observe that the eigenvalues $\mathrm{eig}_n(\boldsymbol{\Psi})$ depend on T_c and $\tau_{c,l}$, which are based on the Poisson process. In general, it is difficult, if not possible, to determine $\mathcal{M}_\eta(s)$ in (6.23). However, for an uncoded multiband system (i.e., when the number of jointly encoded subcarriers is $M = 1$), we can get a closed form.

In case of no coding, the eigenvalue of matrix $\boldsymbol{\Psi}$ in (6.21) is

$$\mathrm{eig}(\boldsymbol{\Psi}) = |d - \hat{d}|^2 \mathrm{eig}\left(\mathbf{W}_p \boldsymbol{\Omega} \mathbf{W}_p^{\mathcal{H}}\right) = |d - \hat{d}|^2, \tag{6.24}$$

in which $(\cdot)^{\mathcal{H}}$ denotes a conjugate transpose operation and the second equality follows from the fact that the matrix

$$\mathbf{W}_p \boldsymbol{\Omega} \mathbf{W}_p^{\mathcal{H}} = \sum_{c=0}^{C} \sum_{l=0}^{L} \Omega_{c,l} = 1. \tag{6.25}$$

By substituting $\mathrm{eig}(\boldsymbol{\Psi}) = |d - \hat{d}|^2$ into (6.23) and then substituting the resulting MGF into (6.15), we arrive at the following result.

Theorem 6.1. When there is no coding across subcarriers, the average PEP of a UWB system employing multiband OFDM is given by

$$P_e = \frac{1}{\pi} \int_0^{\pi/2} \left(1 + \frac{\rho}{4 \sin^2 \theta} |d - \hat{d}|^2\right)^{-1} d\theta \tag{6.26}$$

for any channel model parameters.

The result in Theorem 6.1 shows that the performance of an uncoded multiband OFDM system does not depend on multipath arrival rates or decay factors. In addition, the performance of a UWB system is the same as that of a narrowband system in a Rayleigh fading environment [Sim00]. This implies that we cannot gain from the rich multipath-clustering property of UWB channels if the data are not encoded across subcarriers.

6.2.2 Approximate PEP Formulation

In this subsection we establish a PEP approximation which allows us to obtain insightful understanding of the UWB systems when the information is jointly encoded across subcarriers.

According to [Mat92, p. 29], the quadratic form in a zero-mean Gaussian random vector $\mathbf{x} = [x_1, x_2, \ldots, x_M]^T$ can be represented by a weighted summation of $|v_n|^2$, where v_n are mutually independent standard Gaussian random variables, and the weights are the eigenvalues of the covariance matrix of $\mathbf{x}$. Observe from (6.14) that

$$\eta = (\boldsymbol{\Delta}_p \mathbf{H}_p)^{\mathcal{H}} \boldsymbol{\Delta}_p \mathbf{H}_p \tag{6.27}$$

is in quadratic form and $E[\Delta_p H_p] = 0$. Therefore, using the representation of the quadratic form in [Mat92, p. 29], we can approximate η as

$$\eta \approx \sum_{n=1}^{M} \mathrm{eig}_n(\boldsymbol{\Phi}) |\mu_n|^2, \tag{6.28}$$

where μ_n are iid zero-mean Gaussian random variables with unit variance, and

$$\boldsymbol{\Phi} = E[\Delta_p H_p (\Delta_p H_p)^{\mathcal{H}}] = \Delta_p R_M \Delta_p^{\mathcal{H}}, \tag{6.29}$$

in which

$$R_M = E\left[H_p H_p^{\mathcal{H}}\right] \tag{6.30}$$

is an $M \times M$ correlation matrix. Let the eigenvalues $\mathrm{eig}_n(\boldsymbol{\Phi})$ be arranged in nonincreasing order as

$$\mathrm{eig}_1(\boldsymbol{\Phi}) \geq \mathrm{eig}_2(\boldsymbol{\Phi}) \geq \cdots \geq \mathrm{eig}_M(\boldsymbol{\Phi}). \tag{6.31}$$

By Ostrowski's theorem [Hor85, p. 224], the eigenvalues of $\boldsymbol{\Phi}$ are given by

$$\mathrm{eig}_n(\boldsymbol{\Phi}) = \mathrm{eig}_n(\Delta_p R_M \Delta_p^{\mathcal{H}}) = \nu_n \mathrm{eig}_n(R_M), \tag{6.32}$$

where ν_n is a nonnegative real number that satisfies

$$\mathrm{eig}_M(\Delta_p \Delta_p^{\mathcal{H}}) \leq \nu_n \leq \mathrm{eig}_1(\Delta_p \Delta_p^{\mathcal{H}}) \tag{6.33}$$

for $n = 1, 2, \ldots, M$. From (6.28) and (6.32) we can approximate the MGF in (6.23) as

$$\mathcal{M}_\eta(s) \approx \prod_{n=1}^{M} \frac{1}{1 - s\nu_n \mathrm{eig}_n(R_M)}. \tag{6.34}$$

The remaining problem is to determine the matrix R_M. We observe that the (n, n')th entry of the matrix R_M is $E[H(n)H(n')^*]$ for $0 \leq n, n' \leq M - 1$. The elements on the main diagonal of R_M are given by

$$R(n, n) = E[|H(n)|^2] = \sum_{c=0}^{C} \sum_{l=0}^{L} E[|\alpha(c, l)|^2] = 1. \tag{6.35}$$

The off-diagonal elements of R_M, $R(n, n')$ for $n \neq n'$, can be evaluated as follows:

$$
\begin{aligned}
R(n, n') &= E[H(n)H(n')^*] \\
&= \sum_{c=0}^{C} \sum_{l=0}^{L} E[E[|\alpha(c, l)|^2] \exp(-\mathrm{j}2\pi(n - n')\Delta f(T_c + \tau_{c,l}))] \\
&\triangleq R(n - n').
\end{aligned} \tag{6.36}
$$

Substitute (2.17) into (6.36), resulting in

$$R(m) = \sum_{c=0}^{C} \sum_{l=0}^{L} \Omega_{0,0} G_{c,l}(m), \tag{6.37}$$

where

$$G_{c,l}(m) = E\left[\exp\left(-g\left(\frac{1}{\Gamma}, m\right)T_c - g\left(\frac{1}{\gamma}, m\right)\tau_{c,l}\right)\right] \tag{6.38}$$

and $g(a, m) \triangleq a + j\, 2\pi m\, \Delta f$.

To calculate $G_{c,l}(m)$ in (6.38), we denote x_i as an interarrival time between the clusters i and $i-1$. According to the Poisson distribution of the cluster delays, x_i can be modeled as iid exponential random variables with parameter Λ, and the delay of the cth cluster, T_c, can be expressed as $T^c = \sum_{i=0}^{c} x_i$. Similarly, let $v_{c,j}$ denote an interarrival time between rays j and -1 in the cth cluster. We can also model $v_{c,j}$ as iid exponential random variables with parameter λ, and the delay of the lth path within cluster c can be given by $\tau_{c,l} = \sum_{j=0}^{l} v_{c,j}$. By rewriting $G_{c,l}(m)$ in terms of x_i and $v_{c,j}$, (6.38) can be simplified to

$$G_{c,l}(m) = E\left[\prod_{i=0}^{c}\exp\left(-g\left(\frac{1}{\Gamma}, m\right)x_i\right)\right]E\left[\prod_{j=0}^{l}\exp\left(-g\left(\frac{1}{\gamma}, m\right)v_{c,j}\right)\right]$$
$$= \left(\frac{\Lambda}{\Lambda + g(1/\Gamma, m)}\right)^c \left(\frac{\lambda}{\lambda + g(1/\gamma, m)}\right)^l. \tag{6.39}$$

Substitute (6.39) into (6.37), and use the fact that for UWB channels, the number of clusters C and the number of rays L are generally large. Then we obtain

$$R(m) = \Omega_{0,0}\frac{\Lambda + g(1/\Gamma, m)}{g(1/\Gamma, m)}\frac{\lambda + g(1/\gamma, m)}{g(1/\gamma, m)}. \tag{6.40}$$

Finally, by substituting (6.34) into (6.15), we can characterize the multiband OFDM performance as follows.

Theorem 6.2. When the information is encoded jointly across M ($1 \leq M \leq N$) subcarriers, the average PEP of a multiband OFDM system can be approximated as

$$P_e \approx \frac{1}{\pi}\int_0^{\pi/2}\prod_{n=1}^{M}\left(1 + \frac{\rho v_n}{4\sin^2\theta}\mathrm{eig}_n(\mathbf{R}_M)\right)^{-1}d\theta, \tag{6.41}$$

where the $M \times M$ matrix $\mathbf{R}_M$ is given by

$$\mathbf{R}_M = \begin{pmatrix} 1 & R(1)^* & \cdots & R(M-1)^* \\ R(1) & 1 & \cdots & R(M-2)^* \\ \vdots & \vdots & \ddots & \vdots \\ R(M-1) & R(M-2) & \cdots & 1 \end{pmatrix}_{M \times M}, \tag{6.42}$$

and the $R(m)$ for $m = 1, 2, \ldots, M-1$ are as defined in (6.40).

It is worth noting that the result in Theorem 6.2 can be extended straightforwardly to the case when interleaving or permutation over different subcarriers is applied.

To be specific, if the data matrix is permuted such that the data symbol $d_p(n)$ is transmitted in subcarrier $\sigma_p(n)$ instead of subcarrier n, where n, $\sigma_p(n) \in \{0, 1, \ldots, N-1\}$ [Liu03, Su05a], the PEP performance of the permuted data matrix is of the same form as (6.41) with the off-diagonal elements of matrix $\mathbf{R}_M$ replaced by $R(n, n') = R(\sigma_p(n) - \sigma_p(n'))$.

In the sequel we discuss the PEP approximations in Theorem 6.2 for two special cases to get some insightful understanding.

1. In the case of no coding (i.e., $M = 1$), the correlation matrix in (6.42) becomes $\mathbf{R}_1 = 1$ and $v_1 = |d - \hat{d}|^2$. Thus, the PEP can be obtained from (6.41) as

$$P_e \approx \frac{1}{\pi} \int_0^{\pi/2} \left(1 + \frac{\rho}{4 \sin^2 \theta} |d - \hat{d}|^2\right)^{-1} d\theta, \qquad (6.43)$$

which is consistent with the exact PEP given in (6.26).

2. When the information is encoded jointly across two subcarriers (i.e., $M = 2$), the eigenvalues of matrix $\mathbf{R}_2$ are $1 + |R(1)|$ and $1 - |R(1)|$. Substituting these eigenvalues into (6.41), we obtain the approximate PEP:

$$P_e \approx \frac{1}{\pi} \int_0^{\pi/2} \left(1 + \frac{\rho J + \rho^2 v_1 v_2 (1 - B^2)}{16 \sin^4 \theta}\right)^{-1} d\theta, \qquad (6.44)$$

where $J = 4 \sin^2 \theta [v_1 + v_2 + B(v_1 - v_2)]$,

$$B = \Omega_{0,0} \frac{[(\Lambda + 1/\Gamma)^2 + b]^{1/2} [(\lambda + 1/\gamma)^2 + b]^{1/2}}{[(1/\Gamma)^2 + b]^{1/2} [(1/\gamma)^2 + b]^{1/2}}, \qquad (6.45)$$

and $b = (2\pi \Delta f)^2$. In a UWB system, b is normally much less than $1/\gamma^2$ and $1/\Gamma^2$. Hence, (6.45) can be approximated by

$$B \approx \Omega_{0,0} (\Lambda \Gamma + 1)(\lambda \gamma + 1). \qquad (6.46)$$

Observe that for the uncoded multiband OFDM system, the performance does not depend on the clustering characteristic. However, in case of joint encoding across multiple subcarriers, the PEP in (6.41) reveals that the multiband OFDM performance depends on the correlations in the frequency response among different subcarriers, $R(m)$, which in turn relate to the path arrival rates and decay factors. Specifically, if the number of jointly encoded subcarriers is $M = 2$, the result in ((6.44) brings out that the UWB performance is related to the channel model parameters through the factor B defined in (6.45). This means that the performance of multiband OFDM system depends on both cluster and ray arrival rates as well as their decay factors. In a short-range (0 to 4 m) line-of-sight environment (e.g., scenario for channel model 1 [Foe03b], the product of the cluster arrival rate and cluster decay factor can be much less than 1 ($\Lambda \Gamma \ll 1$). In such a situation, (6.46) can be further simplified to $B \approx \Omega_{0,0}(\lambda \gamma + 1)$, which implies that the performance depends heavily only on the ray arrival rate and ray decay factor. The intuition behind this result is that when both cluster arrival rate and cluster decay factor are small, the effect of the first cluster will dominate. Hence, the performance can be approximated by taking into consideration only the first cluster. On the other hand, when both ray arrival rate and ray decay factor are small such that the product of these two parameters is much less than 1 ($\lambda\gamma \ll 1$), (6.46) reduces to $B \approx \Omega_{0,0}(\Lambda\Gamma + 1)$, which indicates that only the first path in each cluster affects the performance seriously.

For instance, suppose that each data symbol d is transmitted repeatedly in two subcarriers and that channel model parameters follow those specified in the IEEE 802.15.3a channel modeling report [Foe03b]. Let $v = |d - \hat{d}|^2$ and $\Delta f = 4.125$ MHz; then the approximate PEP can be obtained from (6.44) as follows:

- In CM1, $\Lambda = 0.0233$, $\lambda = 2.5$, $\Gamma = 7.1$, $\gamma = 4.3$, and $\Omega_{0,0} = 0.0727$:

$$B = 0.9852 \quad \text{and} \quad P_e \approx \frac{1}{\pi} \int_0^{\pi/2} \left(1 + \frac{0.0294\rho^2 v^2}{16\sin^4\theta}\right)^{-1} d\theta.$$

- In CM4, $\Lambda = 0.0667$, $\lambda = 2.1$, $\Gamma = 24$, $\gamma = 12$, and $\Omega_{0,0} = 0.0147$:

$$B = 0.8351 \quad \text{and} \quad P_e \approx \frac{1}{\pi} \int_0^{\pi/2} \left(1 + \frac{0.3026\rho^2 v^2}{16\sin^4\theta}\right)^{-1} d\theta.$$

We can see from the examples above that UWB performance in CM4 is better than that in CM1. This comes from the fact that the multipath components in CM4 are more random than those in CM1 (as illustrated in Fig. 6.1, which implies that compared with CM1, CM4 has less correlation in the frequency response among different subcarriers and hence yields better performance).

6.2.3 Outage Probability

In this subsection we consider the outage probability analysis for the multiband OFDM system with the S-V fading model. The outage probability [Sim0] is defined as the probability that the combined SNR, ζ, falls below a specified threshold, ζ_o:

$$P_{\text{out}} = P(\zeta \leq \zeta_o) = \int_0^{\zeta_o} p_\zeta(x)\,dx, \tag{6.47}$$

where $p_\zeta(x)$ denotes the PDF of ζ. Since the information is jointly en/decoded for each data matrix $\mathbf{D}_p$, the combined SNR can be defined as

$$\zeta = \frac{E_s\|X(\mathbf{D}_p)\mathbf{H}_p\|^2}{E[\|\mathbf{Z}_p\|^2]} = \frac{\rho}{M}\sum_{n=0}^{M-1}|H_p(n)|^2, \tag{6.48}$$

in which $\rho = E_s/N_0$, as defined previously. Denote

$$\xi = \sum_{n=0}^{M-1}|H_p(n)|^2; \tag{6.49}$$

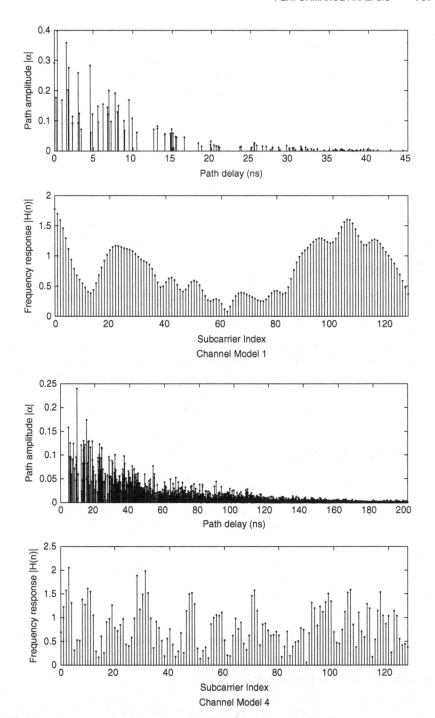

Figure 6.1 One realization of UWB channel generated using the parameters for CM1 and CM4.

then the outage probability can be expressed as

$$P_{\text{out}} = P\left(\xi \le \frac{M\zeta_o}{\rho}\right) = \int_0^{M\zeta_o/\rho} p_\xi(x)\,dx. \tag{6.50}$$

To determine the PDF $p_\xi(x)$, we first obtain the MGF of ξ from the MGF $\mathcal{M}_\eta(s)$ in (6.23) by replacing $\mathbf{\Delta}_p$ with an identity matrix.

According to (6.21) and (6.23), $\mathcal{M}_\xi(s)$ for the case of no coding can be given simply by

$$\mathcal{M}_\xi(s) = \mathrm{E}\left[\left(1 - s\mathbf{W}_p\mathbf{\Omega}\mathbf{W}_p^{\mathcal{H}}\right)^{-1}\right] = (1 - s)^{-1}, \tag{6.51}$$

of which the corresponding PDF is [Sim00, p. 22]

$$p_\xi(x) = \exp(-x); \quad x \ge 0. \tag{6.52}$$

In case of joint encoding across subcarriers, the MGF $\mathcal{M}_\xi(s)$ can be obtained from the approximation approach in Section 6.2.2 as

$$\mathcal{M}_\xi(s) \approx \prod_{n=1}^M \frac{1}{1 - s\,\mathrm{eig}_n(\mathbf{R}_M)} = \sum_{n=1}^M \frac{A_n}{1 - s\,\mathrm{eig}_n(\mathbf{R}_M)}, \tag{6.53}$$

where the equality comes from the technique of partial fractions, $\mathbf{R}_M$ is as defined in (6.42), and A_n is given by

$$A_n = \prod_{n'=1, n' \ne n}^M \frac{\mathrm{eig}_n(\mathbf{R}_M)}{\mathrm{eig}_n(\mathbf{R}_M) - \mathrm{eig}_{n'}(\mathbf{R}_M)}. \tag{6.54}$$

By applying the inverse Laplace transform to the MGF in (6.53), we obtain the PDF of ξ:

$$p_\xi(x) \approx \sum_{n=1}^M \frac{A_n}{\mathrm{eig}_n(\mathbf{R}_M)} \exp\left(-\frac{x}{\mathrm{eig}_n(\mathbf{R}_M)}\right), \quad x \ge 0. \tag{6.55}$$

Finally, substituting the PDF $p_\xi(x)$ above into (6.50) gives rise to the following results.

Theorem 6.3. When there is no coding across subcarriers, the outage probability of a multiband OFDM system is given by

$$P_{\text{out}} = 1 - \exp\left(-\frac{\zeta_o}{\rho}\right) \tag{6.56}$$

for any channel model parameters. When the information is jointly encoded across M ($1 < M \le N$) subcarriers, the outage probability can be approximated as

$$P_{\text{out}} \approx \sum_{n=1}^M A_n\left(1 - \exp\left(-\frac{\zeta_o M}{\rho\,\mathrm{eig}_n(\mathbf{R}_M)}\right)\right), \tag{6.57}$$

where $\mathbf{R}_M$ is as specified in (6.42) and A_n is as defined in (6.54).

From the analysis above, we can see that the outage probability follows the same behaviors as the average PEP. Specifically, the outage probability of an uncoded multiband OFDM system does not depend on the clustering property of a UWB channel, and it is the same as that for narrowband Rayleigh fading environment [Sim00]. When the information is encoded jointly across multiple subcarriers, (6.57) discloses that the outage probability is related to the eigenvalues of the correlation matrix $\mathbf{R}_M$ and depends on the path arrival rates and decay factors.

To gain some insightful understanding on the outage probability formulation in (6.57), let us consider a specific example of jointly encoding across $M = 2$ subcarriers. In this case the outage probability can be approximated as

$$
P_{\text{out}} \approx 1 - 0.5(1 + B^{-1}) \exp\left(-\frac{2\zeta_o}{\rho(1 + B)}\right)
$$

$$
- 0.5(1 - B^{-1}) \exp\left(-\frac{2\zeta_o}{\rho(1 - B)}\right), \tag{6.58}
$$

where B is as defined in (6.45). Since B takes any value between 0 and 1 ($0 < B < 1$), the higher the B, the larger the outage probability P_{out} in (6.58). For instance,

- In CM1, $B = 0.9852$:

$$
P_{\text{out}} \approx 1 - 1.0075 \exp\left(-\frac{1.0075\zeta_o}{\rho}\right) + 0.0075 \exp\left(-\frac{134.91\zeta_o}{\rho}\right).
$$

- In CM4, $B = 0.8351$:

$$
P_{\text{out}} \approx 1 - 1.0987 \exp\left(-\frac{1.0898\zeta_o}{\rho}\right) + 0.0987 \exp\left(-\frac{12.131\zeta_o}{\rho}\right).
$$

From the examples above, we can see that when the SNR ρ is small, $\exp(-2\zeta_o/\rho(1 + B)) \gg \exp(-2\zeta_o/\rho(1 - B))$, and hence the third term in (6.58) is negligible. The outage probability can then be approximated by

$$
P_{\text{out}} \approx 1 - \exp\left(-\frac{\zeta_o}{\rho}\right) \tag{6.59}
$$

for both CM1 and CM4. Such outage probability is the same as that for a narrowband Rayleigh fading channel, which implies that in a low-SNR region we cannot gain from the multipath-clustering property of a UWB channel. As the SNR increases, P_{out} for CM4 drops faster than that for CM1, due to the effect of the third term in P_{out} expressions. Explicitly, the term $0.0987 \exp(-12.131\zeta_o/\rho)$ for CM4 increases with SNR ρ much faster than the term $0.0075 \exp(-134.91\zeta_o/\rho)$ for CM1. Hence, the outage probability performance for CM4 tends to be better than that for CM1 at a high SNR.

6.3 ANALYSIS FOR MIMO MULTIBAND OFDM SYSTEMS

The analysis proposed in Section 6.2 provides a simple but general approach for determining the performances of multiband OFDM systems. In this section we apply the proposed approximation technique to characterize PEP performances of MIMO multiband OFDM system. For convenience, we first briefly describe a MIMO multiband OFDM system model, as discussed in Chapter 5. Then we provide performance analysis of a MIMO multiband OFDM system under a realistic UWB channel model.

6.3.1 MIMO Multiband OFDM System Model

We consider a MIMO multiband OFDM system with N_t transmitting and N_r receiving antennas. The channel impulse response from the ith transmitting antenna to the jth receiving antenna during the kth OFDM block is modeled as

$$h_{ij}^k(t) = \sum_{c=0}^{C} \sum_{l=0}^{L} \alpha_{ij}^k(c, l)\delta(t - T_c - \tau_{c,l}), \tag{6.60}$$

where $\alpha_{ij}^k(c, l)$ is the multipath gain coefficient with $E[|\alpha_{ij}^k(c, l)|^2] = \Omega_{c,l}$. We assume that the average powers $\Omega_{c,l}$ and the delays T_c and $\tau_{c,l}$ are the same for every transmitter–receiver link.

At the transmitter, the information is jointly encoded across N_t transmitting antennas, M OFDM subcarriers, and K OFDM blocks. Each STF codeword can be expressed as an $KM \times N_t$ matrix:

$$\mathbf{D}_p = \left[\left(\mathbf{D}_p^0\right)^{\mathrm{T}} \quad \left(\mathbf{D}_p^1\right)^{\mathrm{T}} \quad \cdots \quad \left(\mathbf{D}_p^{K-1}\right)^{\mathrm{T}} \right]^{\mathrm{T}}, \tag{6.61}$$

where

$$\mathbf{D}_p^k = \left[\mathbf{d}_{p,1}^k \mathbf{d}_{p,2}^k \cdots \mathbf{d}_{p,N_t}^k\right] \tag{6.62}$$

and

$$\mathbf{d}_{p,i}^k = \left[d_i^k(pM)d_i^k(pM + 1)\cdots d_i^k(pM + M - 1)\right]^{\mathrm{T}} \tag{6.63}$$

for $i = 1, 2, \ldots, N_t$ and $k = 0, 1, \ldots, K - 1$. The symbol $d_i^k(n)$, $n = 0, 1, \ldots, N$, represents the complex symbol to be transmitted over subcarrier n by transmitting antenna i during the kth OFDM symbol period. The matrix $\mathbf{D}_p$ is normalized to have average energy $E[\|\mathbf{D}_p\|^2] = KMN_t$. At the kth OFDM block, each vector

$$\mathbf{d}_i^k \triangleq \left[\left(\mathbf{d}_{0,i}^k\right)^{\mathrm{T}} \left(\mathbf{d}_{1,i}^k\right)^{\mathrm{T}} \cdots \left(\mathbf{d}_{P-1,i}^k\right)^{\mathrm{T}} \mathbf{0}_{(N-PM)\times1}\right]^{\mathrm{T}} \tag{6.64}$$

is OFDM-modulated and transmitted by transmitter antenna i.

The signal received at the nth subcarrier at receiver antenna j during the kth OFDM symbol duration can be expressed as

$$y_j^k(n) = \sqrt{\frac{E_s}{N_t}} \sum_{i=1}^{N_t} d_i^k(n)H_{ij}^k(n) + z_j^k(n), \tag{6.65}$$

where

$$H_{ij}^k(n) = \sum_{c=0}^{C} \sum_{l=0}^{L} \alpha_{ij}^k(c, l) \exp[-j2\pi n \Delta f(T_c + \tau_{c,l})] \qquad (6.66)$$

is the frequency response of the channel at subcarrier n between the ith transmitting and the jth receiving antenna during the kth OFDM block, $z_j^k(n)$ is the zero-mean Gaussian noise with variance N_0, and the factor $\sqrt{E_s/N_t}$ guarantees that the average energy per symbol transmitted is E_s, independent of the number of transmitting antennas. We assume that the channel-state information $H_{ij}^k(n)$ is known at the receiver, and the receiver exploits a maximum-likelihood decoder, where the decoding process is performed jointly across N_r receiver antennas.

Due to the band hopping, the K OFDM symbols in each STF codeword are sent over different subbands. With an ideal band hopping, we assume that the signals transmitted over K different frequency bands undergo independent fading. We also assume that the MIMO channel is spatially uncorrelated [i.e., path gains $\alpha_{ij}^k(c, l)$ are independent for different paths and different pairs of transmitting and receiving antennas].

6.3.2 Pairwise Error Probability

Similarly, the PEP between two distinct STF codewords $\mathbf{D}_p$ and $\hat{\mathbf{D}}_p$ can be given by

$$P_e = \mathrm{E}\left[Q\left(\sqrt{\frac{\rho}{2N_t} \sum_{j=1}^{N_r} \|\mathbf{\Delta}_p \mathbf{H}_{p,j}\|^2}\right)\right], \qquad (6.67)$$

where $\mathbf{\Delta}_p = X(\mathbf{D}_p) - X(\hat{\mathbf{D}}_p)$ is a codeword difference matrix in which $X(\mathbf{D}_p)$ converts each column of $\mathbf{D}_p$ into a diagonal matrix and results in an $KM \times KMN_t$ matrix:

$$\begin{aligned} X(\mathbf{D}_p) &= X([\mathbf{d}_{p,1} \cdots \mathbf{d}_{p,N_t}]) \\ &= [\mathrm{diag}(\mathbf{d}_{p,1}) \cdots \mathrm{diag}(\mathbf{d}_{p,N_t})]. \end{aligned} \qquad (6.68)$$

In (6.67), $\mathbf{H}_{p,j}$ is a $KMN_t \times 1$ channel matrix formatted as

$$\mathbf{H}_{p,j} = \left[\mathbf{H}_{p,1j}^{\mathrm{T}} \mathbf{H}_{p,2j}^{\mathrm{T}} \cdots \mathbf{H}_{p,N_t j}^{\mathrm{T}}\right]^{\mathrm{T}}, \qquad (6.69)$$

in which

$$\begin{aligned} \mathbf{H}_{p,ij} = \big[&H_{ij}^0(pM) \cdots H_{ij}^0(pM + M - 1) \cdots \\ &H_{ij}^{K-1}(pM) \cdots H_{ij}^{K-1}(pM + M - 1) \big]^{\mathrm{T}}. \end{aligned} \qquad (6.70)$$

Following the same procedure as in single-antenna transmission, we first obtain

$$\eta = \sum_{j=1}^{N_r} \|\mathbf{\Delta}_p \mathbf{H}_p\|^2 \approx \sum_{j=1}^{N_r} \sum_{n=1}^{M} \mathrm{eig}_n(\mathbf{\Phi}_j)|\mu_{j,n}|^2, \qquad (6.71)$$

where

$$\Phi_j = \Delta_p E\left[\mathbf{H}_{p,j}\mathbf{H}_{p,j}^{\mathcal{H}}\right]\Delta_p^{\mathcal{H}}, \tag{6.72}$$

and $\mu_{j,n}$ are iid zero-mean Gaussian random variables with unit variance. Based on the assumption of independent channels, the matrix $E[\mathbf{H}_{p,j}\mathbf{H}_{p,j}^{\mathcal{H}}]$ can be simplified to

$$E\left[\mathbf{H}_{p,j}\mathbf{H}_{p,j}^{\mathcal{H}}\right] = \mathbf{I}_{KN_t} \otimes \mathbf{R}_M, \tag{6.73}$$

where $\otimes$ denotes the Kronecker product [Hor05], $\mathbf{I}_M$ represents an $M \times M$ identity matrix, and $\mathbf{R}_M$ is specified in (6.42). Therefore, we can rewrite the expression for Φ_j in (6.71) as

$$\Phi_j = (\mathbf{D}_p - \hat{\mathbf{D}}_p)(\mathbf{D}_p - \hat{\mathbf{D}}_p)^{\mathcal{H}} \circ (\mathbf{I}_K \otimes \mathbf{R}_M), \tag{6.74}$$

where $\circ$ denotes the Hadamard [Hor85]. To simplify the notation, we denote

$$\mathbf{S} \triangleq (\mathbf{D}_p - \hat{\mathbf{D}}_p)(\mathbf{D}_p - \hat{\mathbf{D}}_p)^{\mathcal{H}}. \tag{6.75}$$

Finally, substituting (6.74) into (6.71) and using the MGF of η, the average PEP between $\mathbf{D}_p$ and $\hat{\mathbf{D}}_p$ can be approximated as

$$P_e \approx \frac{1}{\pi} \int_0^{\pi/2} \prod_{n=1}^{KN_p} \left(1 + \frac{\rho}{4N_t \sin^2 \theta} \mathrm{eig}_n(\mathbf{S} \circ (\mathbf{I}_K \otimes \mathbf{R}_M))\right)^{-N_r} d\theta. \tag{6.76}$$

From (6.76) it is clear that the multiband MIMO multiband OFDM performance depends on both STF codeword and channel model parameters through the eigenvalues of matrix $\mathbf{S} \circ (\mathbf{I}_K \otimes \mathbf{R}_M)$. If the information is repeated over K OFDM symbols, that is,

$$\mathbf{D}_p^0 = \mathbf{D}_p^1 = \cdots = \mathbf{D}_p^{K-1}, \tag{6.77}$$

the PEP in (6.76) becomes

$$P_e \approx \frac{1}{\pi} \int_0^{\pi/2} \prod_{n=1}^{M} \left(1 + \frac{\rho}{4N_t \sin^2 \theta} \mathrm{eig}_n(\mathbf{S}_0 \circ \mathbf{R}_M)\right)^{-KN_r} d\theta, \tag{6.78}$$

where

$$\mathbf{S}_0 \triangleq (\mathbf{D}_p^0 - \hat{\mathbf{D}}_p^0)(\mathbf{D}_p^0 - \hat{\mathbf{D}}_p^0)^{\mathcal{H}}. \tag{6.79}$$

At high SNR, the approximate PEP in (6.78) can be upper bounded as

$$P_e \lesssim \prod_{n=1}^{r} \left(\frac{\rho}{4N_t} \mathrm{eig}_n(\mathbf{S}_0 \circ \mathbf{R}_M)\right)^{-KN_r}, \tag{6.80}$$

which implies a diversity gain of

$$G_d = rKN_r \tag{6.81}$$

and a coding gain of

$$G_c = \frac{1}{4N_t} \left(\prod_{n=1}^{r} \text{eig}_n(\mathbf{S}_0 \circ \mathbf{R}_M) \right)^{1/r}, \tag{6.82}$$

where r denotes the rank of matrix $\mathbf{S}_0 \circ \mathbf{R}_M$. Since a UWB channel contains a large number of resolvable paths, $\mathbf{R}_M$ is generally of full rank. This leads to the interesting observation that the multiband OFDM system achieves the same diversity advantage in different channel environments. Only the system coding gain depends heavily on the cluster arriving fading paths. To get some insight, we provide a specific example in the following subsection.

6.3.3 Example: Repetition STF Coding Based on Alamouti's Structure

Consider a MIMO multiband OFDM system employing two transmitting antennas and a repetition-coded STF code [Su05c] based on Alamouti's structure [Ala98]. Suppose that the number of jointly encoded subcarriers M is an even integer; then the codeword $\mathbf{D}_p^k$ is given by

$$\mathbf{D}_p^k = \left(\mathbf{I}_2 \otimes \mathbf{1}_{M/2 \times 1} \right) \begin{pmatrix} d_1 & d_2 \\ -d_2^* & d_1^* \end{pmatrix}, \tag{6.83}$$

where $\mathbf{1}_{m \times n}$ denotes an $m \times n$ all-1 matrix, and the d_i's are selected from BPSK or QPSK constellations. Note that $\mathbf{D}_p^k$ is the same for all k's. From the code structure in (6.83), we have

$$\mathbf{S}_0 \circ \mathbf{R}_M = v(\mathbf{I}_2 \otimes \mathbf{1}_{M/2 \times M/2}) \circ \mathbf{R}_M = v\mathbf{I}_2 \otimes \mathbf{R}_{M/2}, \tag{6.84}$$

where $v \triangleq \sum_{i=1}^{2} |d_i - \hat{d}_i|^2$. Substituting (6.78) into (6.78) results in an approximate PEP:

$$P_e \approx \frac{1}{\pi} \int_0^{\pi/2} \prod_{n=1}^{M/2} \left(1 + \frac{\rho v}{8 \sin^2 \theta} \text{eig}_n(\mathbf{R}_{M/2}) \right)^{-2KN_r} d\theta. \tag{6.85}$$

The PEP in (6.85) can easily be obtained for any given values of M. For instance, the PEP expressions for cases of jointly coding across two and four subcarriers are given as follows.

1. For $M = 2$, the approximate PEP is simply

$$P_e \approx \frac{1}{\pi} \int_0^{\pi/2} \left(1 + \frac{\rho v}{8 \sin^2 \theta} \right)^{-2KN_r} d\theta \leq \left(\frac{\rho v}{8} \right)^{-2KN_r},$$

which indicates a diversity gain of $2KN_r$ and a coding gain of $0.125v$, independent of the channel model parameters. The PEP in this case implies that we cannot gain from the multipath-clustering property of a UWB channel.

2. For $M = 4$, the PEP can be approximated as

$$P_e \approx \frac{1}{\pi} \int_0^{\pi/2} \left(1 + \frac{\rho^2 v^2 (1 - B^2)}{64 \sin^4 \theta} \right)^{-2KN_r} d\theta$$

$$\lesssim \left(\frac{\rho v}{8} \sqrt{1 - \Omega_{0,0}^2 (\Lambda\Gamma + 1)^2 (\lambda\gamma + 1)^2} \right)^{-4KN_r}.$$

Clearly, the diversity gain is $4KN_r$ for each channel model, whereas the coding gain is about $0.0214v$ for CM1 and $0.0688v$ for CM4. Such a coding advantage makes the performance of multiband OFDM system under CM4 superior to that under CM1.

The results in this section disclose that regardless of the random-clustering behavior of UWB channels, the diversity gain can be improved by increasing the number of jointly encoded subcarriers, the number of jointly encoded OFDM symbols, or the number of antennas. Nevertheless, increasing the number of jointly encoded subcarriers leads to the loss in coding gain. As shown in the examples above, a diversity order of 4 can be achieved by employing two transmitting and two receiving antennas. The same diversity order can be obtained by employing one receiver antenna but increasing the number of jointly encoded subcarriers from two to four. However, the coding gain is reduced from $0.125v$ to about $0.0214v$ for CM1 and $0.0688v$ for CM4.

6.4 SIMULATION RESULTS

We performed simulations for a multiband OFDM system with $N = 128$ subcarriers and a subband bandwidth of 528 MHz. Each OFDM symbol was of duration $T = 242.42$ ns. After adding the cyclic prefix of length 60.61 ns and the guard interval of length 9.47 ns, the symbol duration became 312.5 ns. The channel model parameters followed those for CM1 and CM4 [Foe03b]. In our simulations, the data matrix $\mathbf{D}$ in (6.3) was constructed via a repetition mapping. For single-antenna transmission, each data matrix $\mathbf{D}_p$ contained only one information symbol d_p (i.e., $\mathbf{D}_p = d_p \cdot \mathbf{1}_{M \times 1}$). The data matrix $\mathbf{D}_p$ for a system with two transmit antennas was constructed according to (6.83).

Figure 6.2(a) and (b) are comparisons between the theoretical PDF of the normalized SNR ξ given in (6.55) and computer simulations for the case of no coding and joint encoding across two subcarriers. There is a good match between the theoretical and simulation results. Figure 6.2(a) confirms the analysis in Section 6.2.3 that for an uncoded system, the PDF of the SNR is the same for different channel environments. Figure 6.2(b) shows that the PDF of the SNR of the coded system depends on the underlying channel model, as expected. Furthermore, Fig. 6.2(b) indicates that the system under CM4 has more opportunity to take on larger SNR values, which implies a better performance than that under CM1.

Figures 6.3 and 6.4 depict the average SER performances of a single-antenna multiband OFDM system as functions of average SNR per bit (E_b/N_0) in decibels.

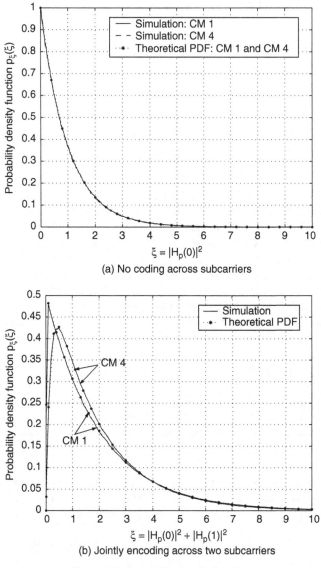

Figure 6.2 Probability density function.

We used BPSK modulation for the performances in Fig. 6.3 and QPSK for those in Fig. 6.4. With BPSK symbols, the average SER is the equivalent of the PEP performance. For QPSK we used the union bound [Sim00] to obtain the average SER from the PEP formulation. In Fig. 6.3(a) and 6.4(a), we show the simulated and theoretical performances of a multiband OFDM system without coding ($M = 1$). We observe that the performances of a UWB system are almost the same in CM1 and CM4, and they

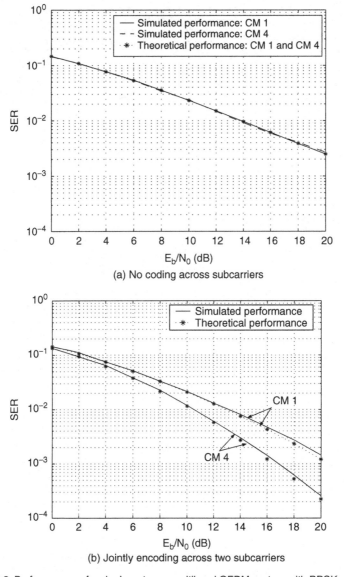

(a) No coding across subcarriers

(b) Jointly encoding across two subcarriers

Figure 6.3 Performances of a single-antenna multiband OFDM system with BPSK symbols.

are close to the exact PEP calculation in (6.26). The simulation results confirm the theoretical expectation that the performances of multiband OFDM systems without coding across subcarriers are the same for every channel environment. Figures 6.3(b) and 6.4(b) show the performances of a multiband OFDM system with the information jointly encoded across two subcarriers ($M = 2$). We can see that the theoretical approximations obtained from (6.41) are close to the performances simulated for

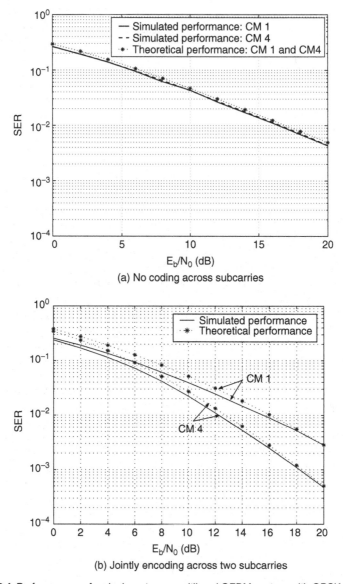

Figure 6.4 Performances of a single-antenna multiband OFDM system with QPSK symbols.

both CM1 and CM4. In addition, the performance obtained under CM4 is superior to that under CM1, which is in agreement with the theoretical results in Section 6.2.2. Figures 6.3(b) and 6.4(b) validate that the PEP approximations can well reflect the multipath-rich and random-clustering characteristics on the performances of UWB systems.

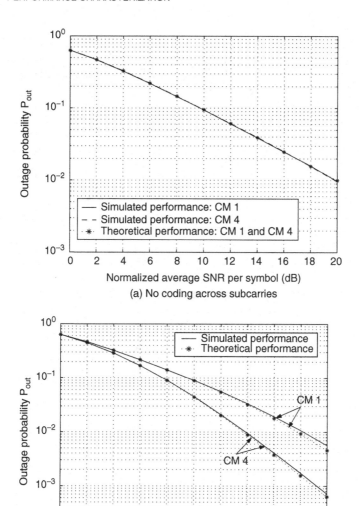

Figure 6.5 Outage probability of a single-antenna multiband OFDM system.

Figure 6.5(a) and (b) plot the outage probability P_{out} versus normalized average SNR ρ/ζ_o in decibels. We can observe that the outage probability follows the same tendencies as the average SER. The uncoded system yields the same outage probability in both CM1 and CM4, whereas the coded system under CM4 achieves a lower outage probability, hence better performance, than that with CM1.

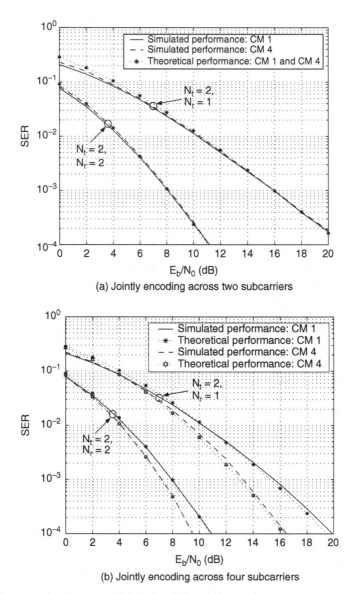

(a) Jointly encoding across two subcarriers

(b) Jointly encoding across four subcarriers

Figure 6.6 Performances of MIMO multiband OFDM system with QPSK symbols.

Figures 6.6(a) and (b) depict the SER performances for a MIMO multiband OFDM system with the information encoded jointly across $N_t = 2$ transmitter antennas, $K = 1$ OFDM symbol, and $M = 2$ and 4 subcarriers. Note that the theoretical SER was obtained from the union bound of the PEP formulation in (6.85). From both figures we can see that the theoretical approximation in (6.85) correctly predicts diversity and coding gains. From Fig. 6.6(b) it is clear that the multiband OFDM system under CM4 outperforms that under CM1, due to the larger coding gain. Figure 6.6 also confirms our observation in Section 6.3 that increasing the number of jointly encoded subcarriers leads to an increase in the diversity gain but a loss in the coding advantage.

6.5 CHAPTER SUMMARY

In this chapter we provide a performance analysis of a multiband OFDM system that captures the unique multipath-rich and random-clustering characteristics of UWB channels. First, exact PEP and outage probability formulations are obtained for the case of no coding across subcarriers. Interestingly, both theoretical and simulation results reveal that the performances of uncoded multiband OFDM systems do not depend on the clustering property. Then we obtain PEP and outage probability approximations in cases when the data are jointly encoded across multiple subcarriers. The theoretical approximations reveal that the performance of a UWB multiband OFDM depends heavily on the correlations in the channel frequency response among different subcarriers, which in turn relate to the cluster arrival rate, the ray arrival rate, and the cluster and ray decay factors. In case of joint coding across two subcarriers, we can draw some interesting conclusion, as follows. When the product of the cluster arrival rate and cluster decay factor is small [e.g., in a short-range (0 to 4 m) line-of-sight scenario], the effect of the first cluster will dominate and the UWB performance can be well approximated by taking into consideration only the first cluster. In contrast, when the product of the ray arrival rate and ray decay factor is much less than 1, the performance depends seriously on only the first path in each cluster. Simulation results confirm that the theoretical analysis can capture successfully the effect of a random-clustering phenomenon on the performance of multiband OFDM system. Finally, we extend the analysis to that for MIMO multiband OFDM systems. It turns out that the coding gain relates strongly to the channel model parameters. On the other hand, the diversity gain can be improved by increasing the number of jointly encoded subcarriers, the number of jointly encoded OFDM symbols, or the number of antennas, regardless of the random-clustering behavior of UWB channels.

7

PERFORMANCE UNDER PRACTICAL CONSIDERATIONS

In previous chapters we analyzed the performance of a multiband OFDM system under perfect frequency and timing synchronization. In the analysis, the channel multipath delays are also assumed to fit inside the cyclic prefix of OFDM symbols, and hence the system does not suffer intersymbol interference (ISI). However, in practice, multipath channel delays can exceed the length of an OFDM cyclic prefix and cause ISI to the received signal in the systems. In addition, the OFDM technique is sensitive to the imperfection of frequency and timing synchronization.

In this chapter we describe performance analysis of multiband OFDM systems that not only captures the characteristics of realistic UWB channels, but also takes into consideration the imperfection of the frequency and timing synchronization and the effect of intersymbol interference [Lai07]. Based on the S-V channel models, we first derive an average SNR of the UWB systems under various synchronization conditions, including perfect synchronization, imperfect timing synchronization, imperfect frequency synchronization, and imperfect frequency and timing synchronization. Then we analyze the multiband OFDM system performance based on the average SNR obtained. We consider two performance metrics: degradation ratio and average BER. Next, we derive a closed-form average BER that provides insightful understanding of the performance of multiband OFDM systems. Then we provide the performance bound of the entire multiband OFDM system. Finally, simulation results validate the theoretical analysis.

Channel and system models are presented in Section 7.1. In Section 7.2, the derivation of the average SNR is presented and the degradation ratio is introduced. The derivation of the average BER is presented in Section 7.3. The performance bound is provided in Section 7.4. In Section 7.5 we present and analyze the numerical and simulation results. Finally, we draw several conclusions in Section 7.6.

Ultra-Wideband Communications Systems: Multiband OFDM Approach, By W. Pam Siriwongpairat and K. J. Ray Liu
Copyright © 2008 John Wiley & Sons, Inc.

Adapted with permission from © 2007 *IEEE International Conference on Acoustics, Speech and Signal Processing*, Vol. 3, Apr. 2007, pp. 111–569-111–572.

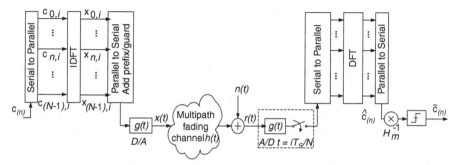

Figure 7.1 System model.

7.1 SYSTEM MODEL

Figure 7.1 illustrates the baseband model of multiband OFDM systems. A data sequence $\{c_{0,i}, c_{1,i}, \ldots, c_{n,i}, \ldots, c_{N-1,i}\}$ with the OFDM symbol index i and the subcarrier index n ($n = 0, 1, \ldots, N - 1$) is input into the systems, where N is the number of subcarriers. The transmitted symbols $c_{n,i}$ are iid with the symbol energy E_s. Since 2 bits form a QPSK symbol in the systems, $E_s = 2E_b$, where E_b is the bit energy. Transmitted OFDM symbols are generated using an N-point inverse discrete Fourier transform (IDFT). The useful OFDM symbols with a duration T_S are preappended by a cyclic prefix (actually, zerotrailing) with a duration T_C to eliminate ISI and appended by a guard interval with a duration T_G to ensure a smooth transition between two consecutive OFDM symbols [Bat03]. The output of the IDFT is

$$x_i(t) = \frac{1}{T_S} \sum_{n=0}^{N-1} c_{n,i} g(t - i T_S') w^{-n(t-iT_S')} \qquad \text{for } -\infty \le i \le \infty, \qquad (7.1)$$

where $T'_S = T_C + T_S + T_G$ is the duration of the OFDM symbol transmitted:

$$g(t) = \begin{cases} 1 & T_C \le t \le T_C + T_S \\ 0 & \text{otherwise} \end{cases} \qquad (7.2)$$

is the rectangular pulse, and $w \triangleq e^{-j2\pi/T_S}$ for notational convenience.

The transmitted signal $x(t) = \sum_{i=-\infty}^{\infty} x_i(t)$ travels through the UWB channel. The received signal $r(t)$ is the sum of the channel output, $y(t)$, and the additive white Gaussian noise (AWGN), $n(t)$:

$$r(t) = y(t) + n(t) = \sum_{i=-\infty}^{\infty} y_i(t) + n(t), \qquad (7.3)$$

where

$$y_i(t) = x_i(t) * h(t)$$

$$= \frac{1}{T_S} \sum_{n=0}^{N-1} c_{n,i} \sum_{l=0}^{L} \sum_{k=0}^{K} \alpha_{k,l} \, g(t - iT'_S - T_l - \tau_{k,l}) w^{-n(t-iT'_S-T_l-\tau_{k,l})} \quad (7.4)$$

is the channel output corresponding to the OFDM symbol x_i, with $*$ denoting convolution. The channel, the transmitted symbols, and the AWGN are assumed mutually independent.

At the receiver, frequency and timing synchronization may not be perfect. The imperfection of frequency synchronization results in a carrier-frequency offset $\Delta f = f_r - f_t$ due to the mismatch between the oscillators of the transmitter and the receiver. Similarly, the error in timing synchronization causes a timing offset τ due to misplacement of the discrete Fourier transform (DFT) window. Figure 7.2 illustrates the imperfection of frequency and timing synchronization. We assume that the cyclic

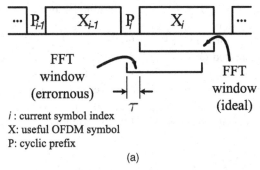

Timing Synchronization Error

i : current symbol index
X: useful OFDM symbol
P: cyclic prefix

(a)

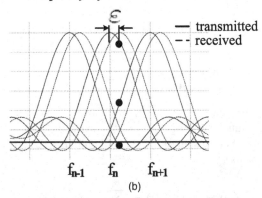

Frequency Synchronization Error

(b)

Figure 7.2 (a) Timing synchronization error; (b) frequency synchronization error.

prefix duration T_C is longer than the length of the timing error [i.e., $\tau \in (-T_C, T_C)$]. Also, we assume that the multipath delay is not longer than the symbol duration (i.e., $T_l + \tau_{k,l} \leq T_S$ for all k, l).

The demodulated signal in subcarrier m during the ith OFDM symbol period can be determined as

$$\hat{c}_{m,i} = \int_{iT_S'+T_C-\tau}^{iT_S'+T_C+T_S-\tau} r(t)e^{-j2\pi(f_{t,m}+\Delta f)(t-iT_S')}\,dt, \tag{7.5}$$

where $f_{t,m}$ is the transmitter carrier frequency corresponding to subcarrier m. Let $\varepsilon = \Delta f/(1/T_S) = \Delta f\, T_S$ be the relative carrier-frequency offset. Substituting (7.3) into (7.5), we have

$$
\begin{aligned}
\hat{c}_{m,i} = &\int_{iT_S'+T_C}^{iT_S'+T_C+T_S} y_i(t-\tau)w^{(m+\varepsilon)(t-iT_S'-\tau)}\,dt \\
&+ \sum_{i'\neq i}\int_{iT_S'+T_C}^{iT_S'+T_C+T_S} y_{i'}(t-\tau)w^{(m+\varepsilon)(t-iT_S'-\tau)}\,dt \\
&+ \int_{iT_S'+T_C}^{iT_S'+T_C+T_S} n(t)w^{(m+\varepsilon)(t-iT_S'-\tau)}\,dt \\
\triangleq\ & A_{m,i} + \hat{c}_{m,i}^{\mathrm{ISI}} + n_{m,i},
\end{aligned}
\tag{7.6}
$$

where $A_{m,i}$ contains information relating to the ith OFDM symbol, $\hat{c}_{m,i}^{\mathrm{ISI}}$ results from the ISI from adjacent OFDM symbols, and $n_{m,i}$ is modeled as a zero-mean complex Gaussian random variable with variance $N_0(n_{m,i} \sim CN(0, N_0))$. Note that the average number of clusters arriving at the receiver at a deterministic time T is $q_0 = \lfloor \Lambda T \rfloor$. For subsequent performance analysis, we assume that all the rays within clusters T_l's whose index $l \leq q_0$ arrive at the receiver before time T.

7.2 AVERAGE SIGNAL-TO-NOISE RATIO

In this section we first derive the expressions of the fading term, the intercarrier interference (ICI) and the ISI, and then determine their variances. Based on the variances obtained, we analyze the UWB system performance in terms of the degradation ratio.

7.2.1 Expressions of Fading Term, ICI, and ISI

As shown in (7.6), the demodulated signal $\hat{c}_{m,i}$ comprises three components, including the signal information $A_{m,i}$, the ISI $\hat{c}_{m,i}^{\mathrm{ISI}}$, and the additive noise $n_{m,i}$. These components are determined as follows. From (7.6) we have

$$A_{m,i} = \int_{iT_S'+T_C}^{iT_S'+T_C+T_S} y_i(t-\tau)w^{(m+\varepsilon)(t-iT_S'-\tau)}\,dt. \tag{7.7}$$

Substituting $y_i(t)$ from (7.4) into (7.7) and applying the change of variable in t, we obtain

$$A_{m,i} = \frac{1}{T_S} \sum_{n=0}^{N-1} c_{n,i} \sum_{l=0}^{L} \sum_{k=0}^{K} \alpha_{k,l} I_{k,l}^c, \tag{7.8}$$

where we define

$$I_{k,l}^c \triangleq \int_{T_C}^{T_C + T_S} g(t - X_{k,l}) w^{-n(t - X_{k,l})} w^{(m+\varepsilon)(t-\tau)} \, dt \tag{7.9}$$

with the rectangular pulse $g(t)$ in (7.2) and $X_{k,l} \triangleq T_l + \tau_{k,l} + \tau$ for notational convenience. Under the assumption that all the rays within clusters T_l's for $l \le q_0 = \lfloor \Lambda T \rfloor$, where T represents deterministic time, arrive at the receiver before T, we have

$$I_{k,l}^c = \begin{cases} I_{k,l}^1 & \text{for } l > l_0 \\ I_{k,l}^2 & \text{for } l \le l_0, \end{cases} \tag{7.10}$$

with $l_0 = \lfloor -\Lambda \tau \rfloor$ when the timing error $\tau < 0$. Note that when the timing error $\tau \ge 0$, (7.10) still holds. In this case, $I_{k,l}^c \equiv I_{k,l}^1$. We are able to show that

$$I_{k,l}^1 = T_S w^{-(n-m-\varepsilon)T_C} w^{-(m+\varepsilon)\tau} \frac{e^{-j2\pi\varepsilon} w^{nX_{k,l}} - w^{(m+\varepsilon)X_{k,l}}}{j2\pi(n-m-\varepsilon)} \tag{7.11}$$

$$I_{k,l}^2 = T_S w^{-(n-m-\varepsilon)T_C} w^{-(m+\varepsilon)\tau} \frac{e^{-j2\pi\varepsilon} w^{(m+\varepsilon)X_{k,l}} - w^{nX_{k,l}}}{j2\pi(n-m-\varepsilon)}. \tag{7.12}$$

Using (7.11) and (7.12) in (7.8), we can express

$$A_{m,i} = c_{m,i} H_m + \hat{c}_{m,i}^{\text{ICI}}, \tag{7.13}$$

where

$$H_m = w^{\varepsilon T_C} w^{-(m+\varepsilon)\tau} \left[\sum_{l=0}^{l_0} \sum_{k=0}^{K} \alpha_{k,l} w^{mX_{k,l}} U_1 + \sum_{l=l_0+1}^{L} \sum_{k=0}^{K} \alpha_{k,l} w^{mX_{k,l}} U_2 \right] \tag{7.14}$$

contains the effect of fading (referred to as the *fading term*), and

$$\hat{c}_{m,i}^{\text{ICI}} = \sum_{n \ne m} c_{n,i} w^{-(n-m-\varepsilon)T_C} w^{-(m+\varepsilon)\tau} \left[\sum_{l=0}^{l_0} \sum_{k=0}^{K} \alpha_{k,l} U_3 + \sum_{l=l_0+1}^{L} \sum_{k=0}^{K} \alpha_{k,l} U_4 \right] \tag{7.15}$$

is ICI from other subcarriers. When $\tau > 0$, l_0 is negative. In such a case, l_0 in the second summation will take a value of zero. Also in (7.14) and (7.15), U_1, U_2, U_3, and U_4 are defined as follows. When ε is not an integer,

$$U_1 = \frac{e^{-j2\pi\varepsilon} w^{\varepsilon X_{k,l}} - 1}{-j2\pi\varepsilon} \tag{7.16}$$

$$U_2 = \frac{e^{-j2\pi\varepsilon} - w^{\varepsilon X_{k,l}}}{-j2\pi\varepsilon} \tag{7.17}$$

$$U_3 = \frac{e^{-j2\pi\varepsilon} w^{(m+\varepsilon)X_{k,l}} - w^{nX_{k,l}}}{j2\pi(n-m-\varepsilon)} \tag{7.18}$$

$$U_4 = \frac{e^{-j2\pi\varepsilon} w^{nX_{k,l}} - w^{(m+\varepsilon)X_{k,l}}}{j2\pi(n-m-\varepsilon)}. \tag{7.19}$$

When $\varepsilon = 0$ (i.e., perfect synchronization), we have

$$U_1 = \lim_{\varepsilon \to 0} \frac{e^{-j2\pi\varepsilon} w^{\varepsilon X_{k,l}} - 1}{-j2\pi\varepsilon} = T_S + X_{k,l} \tag{7.20}$$

$$U_2 = \lim_{\varepsilon \to 0} \frac{e^{-j2\pi\varepsilon} - w^{\varepsilon X_{k,l}}}{-j2\pi\varepsilon} = T_S - X_{k,l}. \tag{7.21}$$

When ε is an integer (i.e., imperfect frequency synchronization with a multiple-subcarrier offset), we have

$$U_3 = \lim_{n \to (m+\varepsilon)} \frac{e^{-j2\pi\varepsilon} w^{(m+\varepsilon)X_{k,l}} - w^{nX_{k,l}}}{j2\pi(n-m-\varepsilon)} = \frac{1}{T_S} w^{(m+\varepsilon)X_{k,l}}(T_S + X_{k,l}) \tag{7.22}$$

$$U_4 = \lim_{n \to (m+\varepsilon)} \frac{e^{-j2\pi\varepsilon} w^{nX_{k,l}} - w^{(m+\varepsilon)X_{k,l}}}{j2\pi(n-m-\varepsilon)} = \frac{1}{T_S} w^{(m+\varepsilon)X_{k,l}}(T_S - X_{k,l}). \tag{7.23}$$

Next, we determine the ISI component, $\hat{c}_{m,i}^{\text{ISI}}$, as follows. From (7.6),

$$\hat{c}_{m,i}^{\text{ISI}} = \sum_{i' \neq i} \int_{iT_S'+T_C}^{iT_S'+T_C+T_S} y_{i'}(t-\tau) w^{(m+\varepsilon)(t-iT_S'-\tau)} \, dt. \tag{7.24}$$

Under the assumptions that $\tau \in (-T_C, T_C)$ and $T_l + \tau_{k,l} \leq T_S$ for all k and l, only the previous $(i-1)$th OFDM symbol involves in the current ith OFDM symbol. Therefore, (7.24) can be simplified to

$$\hat{c}_{m,i}^{\text{ISI}} = \int_{iT_S'+T_C}^{iT_S'+T_C+T_S} y_{i-1}(t-\tau) w^{(m+\varepsilon)(t-iT_S'-\tau)} \, dt. \tag{7.25}$$

Substituting $y_{i-1}(t)$ from (7.4) into (7.25), and applying the change of variable in t, we have

$$\hat{c}_{m,i}^{\text{ISI}} = \frac{1}{T_S} \sum_{n=0}^{N-1} c_{n,i-1} \sum_{l=1}^{L} \sum_{k=1}^{K} \alpha_{k,l} I_{k,l}^s, \tag{7.26}$$

where we define

$$
I_{k,l}^s \triangleq \int_{T_C}^{T_C+T_S} g(t - X_{k,l} + T_S')w^{-n(t-X_{k,l}+T_C+T_G)}w^{(m+\varepsilon)(t-\tau)}\,dt \tag{7.27}
$$

with the rectangular pulse $g(t)$ in (7.2). Based on the assumptions that $\tau \in (-T_C, T_C)$ and $T_l + \tau_{k,l} \leq T_S$ for all k and l, we are able to show that

$$
I_{k,l}^s = T_S w^{-(n-m-\varepsilon)T_C} w^{-(m+\varepsilon)\tau} U_5 \tag{7.28}
$$

for $l > p_0$, where $p_0 = \lfloor \Lambda(T_G + T_C - \tau) \rfloor$. In (7.28),

$$
U_5 = \frac{w^{(m+\varepsilon)(X_{k,l}-T_C-T_G)} - w^{n(X_{k,l}-T_C-T_G)}}{j2\pi(n - m - \varepsilon)} \tag{7.29}
$$

when ε is not an integer, and

$$
\begin{aligned}
U_5 &= \lim_{n \to (m+\varepsilon)} \frac{w^{(m+\varepsilon)(X_{k,l}-T_C-T_G)} - w^{n(X_{k,l}-T_C-T_G)}}{j2\pi(n - m - \varepsilon)} \\
&= w^{(m+\varepsilon)(X_{k,l}-T_C-T_G)}(X_{k,l} - T_C - T_G)
\end{aligned} \tag{7.30}
$$

otherwise. As a result, the ISI component is

$$
\hat{c}_{m,i}^{\text{ISI}} = \sum_{n=0}^{N-1} c_{n,i-1} w^{-(n-m-\varepsilon)T_C} w^{-(m+\varepsilon)\tau} \sum_{l=p_0+1}^{L} \sum_{k=0}^{K} \alpha_{k,l} U_5. \tag{7.31}
$$

7.2.2 Variances of Fading Term, ICI, and ISI

Let us denote the variances of the fading term, the ICI, and the ISI as σ_H^2, σ_C^2, and σ_S^2, respectively. Because the transmitted symbols $c_{n,i}$'s and the multipath gain coefficients $\alpha_{k,l}$'s are zeromean, H_m, $\hat{c}_{m,i}^{\text{ICI}}$, and $\hat{c}_{m,i}^{\text{ISI}}$ are also zeromean. Thus, $\sigma_H^2 = E\{|H_m|^2\}$, $\sigma_C^2 = E\{|\hat{c}_{m,i}^{\text{ICI}}|^2\}$, and $\sigma_H^2 = E\{|\hat{c}_{m,i}^{\text{ISI}}|^2\}$. From the expressions of H_m, $\hat{c}_{m,i}^{\text{ICI}}$, and $\hat{c}_{m,i}^{\text{ISI}}$ derived in Section 7.1, their variances can be determined as follows:

$$
\begin{aligned}
\sigma_H^2 &= \frac{1}{4\pi^2\varepsilon^2} \sum_{l=0}^{l_0} \sum_{k=0}^{K} E\{\Omega_{0,0}e^{-T_l/\Gamma-\tau_{k,l}/\gamma} \\
&\quad \times [2 - (e^{-j2\pi\varepsilon}w^{\varepsilon X_{k,l}} + e^{j2\pi\varepsilon}w^{-\varepsilon X_{k,l}})]\} \\
&\quad + \frac{1}{4\pi^2\varepsilon^2} \sum_{l=l_0+1}^{L} \sum_{k=0}^{K} E\{\Omega_{0,0}e^{-T_l/\Gamma-\tau_{k,l}/\gamma} \\
&\quad \times [2 - (e^{-j2\pi\varepsilon}w^{-\varepsilon X_{k,l}} + e^{j2\pi\varepsilon}w^{\varepsilon X_{k,l}})]\}
\end{aligned} \tag{7.32}
$$

$$
\sigma_C^2 = E_s \sum_{n \neq m} \frac{1}{4\pi^2(n-m-\varepsilon)^2} \sum_{l=0}^{l_0} \sum_{k=0}^{K} E\left\{\Omega_{0,0} e^{-T_l/\Gamma - \tau_{k,l}/\gamma}\right.
$$

$$
\times \left[2 - \left(e^{-j2\pi\varepsilon} w^{-(n-m-\varepsilon)X_{k,l}} + e^{j2\pi\varepsilon} w^{(n-m-\varepsilon)X_{k,l}}\right)\right]\right\}
$$

$$
+ E_s \sum_{n \neq m} \frac{1}{4\pi^2(n-m-\varepsilon)^2} \sum_{l=l_0+1}^{L} \sum_{k=0}^{K} E\left\{\Omega_{0,0} e^{-T_l/\Gamma - \tau_{k,l}/\gamma}\right.
$$

$$
\times \left[2 - \left(e^{-j2\pi\varepsilon} w^{(n-m-\varepsilon)X_{k,l}} + e^{j2\pi\varepsilon} w^{-(n-m-\varepsilon)X_{k,l}}\right)\right]\right\} \tag{7.33}
$$

$$
\sigma_S^2 = E_s \sum_{n=0}^{N-1} \frac{1}{4\pi^2(n-m-\varepsilon)^2} \sum_{l=p_0+1}^{L} \sum_{k=0}^{K} E\left\{\Omega_{0,0} e^{-T_l/\Gamma - \tau_{k,l}/\gamma}\right.
$$

$$
\times \left[2 - \left(w^{(n-m-\varepsilon)(X_{k,l}-T_C-T_G)} + w^{-(n-m-\varepsilon)(X_{k,l}-T_C-T_G)}\right)\right]\right\}. \tag{7.34}
$$

For notational convenience, we introduce the following quantities:

$$
A_1 \triangleq \sum_{l=l_0+1}^{\infty} \sum_{k=1}^{\infty} E\left\{\Omega_{0,0} e^{-T_l/\Gamma - \tau_{k,l}/\gamma} (T_l + \tau_{k,l} - T)^2\right\} \tag{7.35}
$$

$$
A_2 \triangleq \sum_{p=p_0+1}^{\infty} E\left\{\Omega_{0,0} e^{-X_p/\gamma_X} (X_p - T)^2\right\} \tag{7.36}
$$

$$
B_1 \triangleq \sum_{l=l_0+1}^{\infty} \sum_{k=1}^{\infty} E\left\{\Omega_{0,0} e^{-T_l/\Gamma - \tau_{k,l}/\gamma}\left[2 - \left(e^{-j2\pi\varepsilon'} w^{(n-m-\varepsilon)(T_l+\tau_{k,l}-T)}\right.\right.\right.
$$

$$
\left.\left.\left. + e^{j2\pi\varepsilon'} w^{-(n-m-\varepsilon)(T_l+\tau_{k,l}-T)}\right)\right]\right\} \tag{7.37}
$$

$$
B_2 \triangleq \sum_{p=p_0+1}^{\infty} E\left\{\Omega_{0,0} e^{-X_p/\gamma_X}\left[2 - \left(e^{-j2\pi\varepsilon'} w^{(n-m-\varepsilon)(X_p-T)}\right.\right.\right.
$$

$$
\left.\left.\left. + e^{j2\pi\varepsilon'} w^{-(n-m-\varepsilon)(X_p-T)}\right)\right]\right\}, \tag{7.38}
$$

where $E\{\cdot\}$ denotes expectation and $w = e^{-j2\pi/T_S}$, as defined in Section 7.2.1. In the equations above, T_l, $\tau_{k,l}$, and X_p are arrival times in Poisson processes whose rates are Λ, λ, and λ_X and the decay factors are Γ, γ, and γ_X, respectively. X_p can take T_l or $\tau_{k,l}$, depending on particular cases, and will be clarified later. The variances of interest will be separated in terms of A_1, A_2, B_1, and B_2 with different values of T, n, m, ε, ε', λ_X, and γ_X. Notice that T_l, $\tau_{k,l}$, and X_p are l-, k-, and p-Erlang random variables. The derivation of these quantities is presented in the Appendix, and the results follow:

$$
A_1 = \Omega_{0,0}[\Gamma^2 f_3(l_0, \Lambda, \Gamma) f_1(0, \lambda, \gamma) + \gamma^2 f_1(l_0, \Lambda, \Gamma) f_3(0, \lambda, \gamma)
$$

$$
+ T^2 f_1(l_0, \Lambda, \Gamma) f_1(0, \lambda, \gamma) + 2\Gamma\gamma f_2(l_0, \Lambda, \Gamma) f_2(0, \lambda, \gamma)
$$

$$
- 2T \Gamma f_2(l_0, \Lambda, \Gamma) f_1(0, \lambda, \gamma) - 2T \gamma f_1(l_0, \Lambda, \Gamma) f_2(0, \lambda, \gamma)] \tag{7.39}
$$

$$A_2 = \Omega_{0,0} \left[\gamma_X^2 f_3(p_0, \lambda_X, \gamma_X) + T^2 f_1(p_0, \lambda_X, \gamma_X) - 2T\gamma_X f_2(p_0, \lambda_X, \gamma_X) \right] \quad (7.40)$$

$$B_1 = 2\Omega_{0,0} f_1(l_0, \Lambda, \Gamma) f_1(0, \lambda, \lambda) - 2\Omega_{0,0}\beta_T^{l_0+1}\beta_\tau$$

$$\times \left[\frac{\cos\left((l_0+1)\theta_T + \theta_\tau - (2\pi(n-m-\varepsilon)T/T_S) + 2\pi\varepsilon'\right)}{\left(1 + \beta_T^2 - 2\beta_T\cos\theta_T\right)\left(1 + \beta_\tau^2 - 2\beta_\tau\cos\theta_\tau\right)} \right.$$

$$- \frac{\beta_\tau\cos\left((l_0+1)\theta_T - (2\pi(n-m-\varepsilon)T/T_S) + 2\pi\varepsilon'\right)}{\left(1 + \beta_T^2 - 2\beta_T\cos\theta_T\right)\left(1 + \beta_\tau^2 - 2\beta_\tau\cos\theta_\tau\right)}$$

$$- \frac{\beta_T\cos\left(l_0\theta_T + \theta_\tau - (2\pi(n-m-\varepsilon)T/T_S) + 2\pi\varepsilon'\right)}{\left(1 + \beta_T^2 - 2\beta_T\cos\theta_T\right)\left(1 + \beta_\tau^2 - 2\beta_\tau\cos\theta_\tau\right)}$$

$$\left. + \frac{\beta_T\beta_\tau\cos\left(l_0\theta_T - (2\pi(n-m-\varepsilon)T/T_S) + 2\pi\varepsilon'\right)}{\left(1 + \beta_T^2 - 2\beta_T\cos\theta_T\right)\left(1 + \beta_\tau^2 - 2\beta_\tau\cos\theta_\tau\right)} \right] \quad (7.41)$$

$$B_2 = 2\Omega_{0,0} f_1(p_0, \lambda_X, \gamma_X) - 2\Omega_{0,0}\beta_X^{p_0+1}$$

$$\times \left[\frac{\cos\left((p_0+1)\theta_X - (2\pi(n-m-\varepsilon)T/T_S) + 2\pi\varepsilon'\right)}{1 + \beta_X^2 - 2\beta_X\cos\theta_X} \right.$$

$$\left. - \beta_X \frac{\cos\left(p_0\theta_X - (2\pi(n-m-\varepsilon)T/T_S) + 2\pi\varepsilon'\right)}{1 + \beta_X^2 - 2\beta_X\cos\theta_X} \right], \quad (7.42)$$

where

$$f_1(q, a, b) \triangleq \frac{(ab)^{q+1}(ab+1)}{(ab+1)^{p+1}} \quad (7.43)$$

$$f_2(q, a, b) \triangleq \frac{(ab)^{q+1}(ab+q+1)}{(ab+1)^{q+1}} \quad (7.44)$$

$$f_3(q, a, b) \triangleq \frac{(ab)^{q+1}\left[2ab(ab+1) + 2(q+1)ab + (q+1)(q+2)\right]}{(ab+1)^{q+2}} \quad (7.45)$$

and $\beta_\tau, \beta_T, \beta_X, \theta_\tau, \theta_T$, and θ_X follow:

$$\beta_Y \triangleq \frac{\lambda_Y}{\sqrt{(\lambda_Y + 1/\gamma_Y)^2 + 4\pi^2(n-m-\varepsilon)^2/T_S^2}} \quad (7.46)$$

$$\theta_Y \triangleq \arctan\left(\frac{2\pi(n-m-\varepsilon)}{T_S} \frac{\gamma_Y}{\lambda_Y\gamma_Y + 1}\right). \quad (7.47)$$

Now the variances σ_H^2, σ_C^2, and σ_S^2 can be derived in terms of A_1, A_2, B_1, and B_2 as follows. To simplify the presentation, we summarize below the results in case of LOS channels and $\tau < 0$:

1. Perfect frequency and timing synchronization (i.e., $\varepsilon = 0$ and $\tau = 0$):

$$\sigma_H^2 = \Omega_{0,0} + \frac{1}{T_S^2} A_2(T_S, 0, \lambda, \gamma) + \frac{1}{T_S^2} A_2(T_S, 0, \Lambda, \Gamma)$$

$$+ \frac{1}{T_S^2} A_1(T_S, 0, \Lambda, \Gamma, \lambda, \gamma) \tag{7.48}$$

$$\sigma_C^2 = E_s \sum_{n \neq m} \frac{1}{4\pi^2 (n-m)^2} [B_2(0, n, m, 0, 0, 0, \lambda, \gamma)$$

$$+ B_2(0, n, m, 0, 0, 0, \Lambda, \Gamma) + B_1(0, n, m, 0, 0, 0, \Lambda, \Gamma, \lambda, \gamma)] \tag{7.49}$$

$$\sigma_S^2 = E_s \left[\frac{1}{T_S^2} A_2(T_C + T_G, p_0, \Lambda, \Gamma) + \frac{1}{T_S^2} A_1(T_C + T_G, p_0) \right.$$

$$+ \sum_{n \neq m} \frac{1}{4\pi^2 (n-m)^2} [B_2(T_C + T_G, n, m, 0, 0, p_0, \Lambda, \Gamma)$$

$$\left. + B_1(T_C + T_G, n, m, 0, 0, p_0, \Lambda, \Gamma, \lambda, \gamma)] \right] \tag{7.50}$$

2. Imperfect timing synchronization (i.e., $\varepsilon = 0$ and $\tau \neq 0$):

$$\sigma_H^2 = \Omega_{0,0} \frac{(\tau + T_S)^2}{T_S^2} + \frac{1}{T_S^2} A_2(-T_S - \tau, 0, \lambda, \gamma)$$

$$+ \frac{1}{T_S^2} A_2(-T_S - \tau, 0, \Lambda, \Gamma) - \frac{1}{T_S^2} A_2(-T_S - \tau, l_0, \Lambda, \Gamma)$$

$$+ \frac{1}{T_S^2} A_2(T_S - \tau, l_0, \Lambda, \Gamma) + \frac{1}{T_S^2} A_1(-T_S - \tau, 0, \Lambda, \Gamma, \lambda, \gamma)$$

$$- \frac{1}{T_S^2} A_1(-T_S - \tau, l_0, \Lambda, \Gamma, \lambda, \gamma) + \frac{1}{T_S^2} A_1(T_S - \tau, l_0, \Lambda, \Gamma, \lambda, \gamma) \tag{7.51}$$

$$\sigma_C^2 = E_s \sum_{n \neq m} \frac{1}{4\pi^2 (n-m)^2} \left[2\Omega_{0,0} \left(1 - \cos \frac{2\pi (n-m)\tau}{T_S} \right) \right.$$

$$+ B_2(-\tau, n, m, 0, 0, 0, \lambda, \gamma) + B_2(-\tau, n, m, 0, 0, 0, \Lambda, \Gamma)$$

$$\left. + B_1(-\tau, n, m, 0, 0, 0, \Lambda, \Gamma, \lambda, \gamma)] \right. \tag{7.52}$$

$$\sigma_S^2 = E_s \left[\frac{1}{T_S^2} A_2(T_C + T_G - \tau, p_0, \Lambda, \Gamma) + \frac{1}{T_S^2} A_1(T_C + T_G - \tau, p_0, \Lambda, \Gamma, \lambda, \gamma) \right.$$

$$+ \sum_{n \neq m} \frac{1}{4\pi^2(n-m)^2} \left[B_2(T_C + T_G - \tau, n, m, 0, 0, p_0, \Lambda, \Gamma) \right.$$

$$\left. \left. + B_1(T_C + T_G - \tau, n, m, 0, 0, p_0, \Lambda, \Gamma, \lambda, \gamma) \right] \right] \tag{7.53}$$

3. Imperfect frequency synchronization (i.e., $\varepsilon \neq 0$ and $\tau = 0$):

$$\sigma_H^2 = \frac{1}{4\pi^2\varepsilon^2} \left[2\Omega_{0,0}(1 - \cos 2\pi\varepsilon) + B_2(0, 0, 0, \varepsilon, \varepsilon, 0, \lambda, \gamma) \right.$$

$$\left. + B_2(0, 0, 0, \varepsilon, \varepsilon, 0, \Lambda, \Gamma) + B_1(0, 0, 0, \varepsilon, \varepsilon, 0, \Lambda, \Gamma, \lambda, \gamma) \right] \tag{7.54}$$

$$\sigma_C^2 = E_s \sum_{n \neq m} \frac{1}{4\pi^2(n-m-\varepsilon)^2} \left[2\Omega_{0,0}(1 - \cos 2\pi\varepsilon) + B_2(0, n, m, \varepsilon, \varepsilon, 0, \lambda, \gamma) \right.$$

$$\left. + B_2(0, n, m, \varepsilon, \varepsilon, 0, \Lambda, \Gamma) + B_1(0, n, m, \varepsilon, \varepsilon, 0, \Lambda, \Gamma, \lambda, \gamma) \right] \tag{7.55}$$

$$\sigma_S^2 = E_s \sum_{n=0}^{N-1} \frac{1}{4\pi^2(n-m-\varepsilon)^2} \left[B_2(T_C + T_G, n, m, \varepsilon, 0, p_0, \Lambda, \Gamma) \right.$$

$$\left. + B_1(T_C + T_G, n, m, \varepsilon, 0, p_0, \Lambda, \Gamma, \lambda, \gamma) \right] \tag{7.56}$$

4. Imperfect frequency and timing synchronization (i.e., $\varepsilon \neq 0$ and $\tau \neq 0$):

$$\sigma_H^2 = \frac{1}{4\pi^2\varepsilon^2} \left[2\Omega_{0,0} \left(1 - \cos\left(\frac{2\pi\varepsilon\tau}{T_S} + 2\pi\varepsilon \right) \right) + B_2(-\tau, 0, 0, \varepsilon, -\varepsilon, 0, \lambda, \gamma) \right.$$

$$+ B_2(-\tau, 0, 0, \varepsilon, -\varepsilon, 0, \Lambda, \Gamma) - B_2(-\tau, 0, 0, \varepsilon, -\varepsilon, l_0, \Lambda, \Gamma)$$

$$+ B_2(-\tau, 0, 0, \varepsilon, \varepsilon, l_0, \Lambda, \Gamma) + B_1(-\tau, 0, 0, \varepsilon, -\varepsilon, 0, \Lambda, \Gamma, \lambda, \gamma)$$

$$\left. - B_1(-\tau, 0, 0, \varepsilon, -\varepsilon, l_0, \Lambda, \Gamma, \lambda, \gamma) + B_1(-\tau, 0, 0, \varepsilon, \varepsilon, l_0, \Lambda, \Gamma, \lambda, \gamma) \right] \tag{7.57}$$

$$\sigma_C^2 = E_s \sum_{n \neq m} \frac{1}{4\pi^2(n-m-\varepsilon)^2} \left[2\Omega_{0,0} \left(1 - \cos\left(\frac{2\pi(n-m-\varepsilon)\tau}{T_S} - 2\pi\varepsilon \right) \right) \right.$$

$$+ B_2(-\tau, n, m, \varepsilon, -\varepsilon, 0, \lambda, \gamma) + B_2(-\tau, n, m, \varepsilon, -\varepsilon, 0, \Lambda, \Gamma)$$

$$- B_2(-\tau, n, m, \varepsilon, -\varepsilon, l_0, \Lambda, \Gamma) + B_2(-\tau, n, m, \varepsilon, \varepsilon, l_0, \Lambda, \Gamma)$$

$$+ B_1(-\tau, n, m, \varepsilon, -\varepsilon, 0, \Lambda, \Gamma, \lambda, \gamma) - B_1(-\tau, n, m, \varepsilon, -\varepsilon, l_0, \Lambda, \Gamma, \lambda, \gamma)$$

$$\left. + B_1(-\tau, n, m, \varepsilon, \varepsilon, l_0, \Lambda, \Gamma, \lambda, \gamma \right\} \tag{7.58}$$

$$\sigma_S^2 = E_s \sum_{n=0}^{N-1} \frac{1}{4\pi^2(n - m - \varepsilon)^2} [B_2(T_C + T_G - \tau, n, m, \varepsilon, 0, p_0, \Lambda, \Gamma)$$

$$+ B_1(T_C + T_G - \tau, n, m, \varepsilon, 0, p_0, \Lambda, \Gamma, \lambda, \gamma)], \tag{7.59}$$

where $l_0 = \lfloor -\Lambda\tau \rfloor$ and $p_0 = \lfloor \Lambda(T_G + T_C - \tau) \rfloor$. Note that the variances in cases of non-LOS or $\tau \geq 0$ can be obtained simply in a similar way.

7.2.3 Average Signal-to-Noise Ratio and Performance Degradation

In Section 7.2.2, the variances of the fading term, the ICI and the ISI, are obtained. Beside the fading and the interferences, the symbol received is also affected by AWGN $n_{m,i}$, whose variance is N_0. Thus, the average SNR per QPSK symbol can be defined as

$$\overline{\gamma_s}(\varepsilon, \tau) \triangleq \frac{E_s \sigma_H^2}{\sigma_S^2 + \sigma_S^2 + N_0}. \tag{7.60}$$

Since the energy per bit $E_b = 1/2E_s$, the average SNR per bit $\overline{\gamma_b}(\varepsilon, \tau) = 1/2\overline{\gamma_s}(\varepsilon, \tau)$. From the average SNR, the degradation ratio can be defined as

$$D(\varepsilon, \tau) = 10 \log \frac{\overline{\gamma_b}(0, 0)}{\overline{\gamma_b}(\varepsilon, \tau)} \text{ dB}. \tag{7.61}$$

The degradation ratio in (7.61) measures the relative performance of systems: system performance in the imperfect synchronization in comparison with system performance in the perfect synchronization. In Section 7.5 we present and analyze the performance degradation of the UWB systems based on the numerical results obtained from (7.61).

7.3 AVERAGE BIT ERROR RATE

Average BER, which measures absolute performance, is defined as a ratio of the number of bits received incorrectly to the total number of bits sent. According to (7.6) and (7.13), the demodulated signal $\hat{c}_{m,i}$ at the mth subcarrier can be expressed as

$$\hat{c}_{m,i} = c_{m,i} H_m + z_m, \tag{7.62}$$

where $z_m \triangleq \hat{c}_{m,i}^{ICI} + \hat{c}_{m,i}^{ISI} + n_{m,i}$ represents the summation of the ICI, the ISI, and the AWGN. In Section 7.2.1 we have seen that $\hat{c}_{m,i}^{ICI}$ and $\hat{c}_{m,i}^{ISI}$ are the sums of independent but not identically distributed random variables. Thus, we cannot apply the central limit theorem, which requires the summation of iid random variables [Leo94], to model the ICI and the ISI as Gaussian random variables. However, to obtain the performance bound, we model the ICI and the ISI as Gaussian random variables

TABLE 7.1   Data-Rate Modes

Data-Rate Modes	Data Rates (Mbps)	Frequency Spreading Gain	Time Spreading Gain	Overall Spreading Gain
Low rate	53.3–80	2	2	4
Middle rate	106.7–200	1	2	2
High rate	320–480	1	1	1

whose mean is zero and variance is σ_C^2 and σ_S^2, respectively. This can be done because independent Gaussian noise yields the smallest capacity among additive noise processes with fixed variance and mean [Cov91]. Consequently, z_m will be modeled as iid complex Gaussian random variables whose mean is zero and variance is $\sigma^2 z = \sigma^2 c + \sigma^2 s + N_0$.

As presented in Table 4.2, the multiband OFDM system is able to support 10 data rates ranging from 53.3 to 480 Mbps [Bat03]. These data rates can be grouped into three data-rate modes according to three different overall spreading gains, as summarized in Table 7.1. All three modes have the same received signal model:

$$\hat{\mathbf{c}} = c_{m,i}\mathbf{h} + \mathbf{z}, \tag{7.63}$$

where $\hat{\mathbf{c}}$ is a vector comprising demodulated signals $\hat{c}_{m,i}$, $\mathbf{h}$ is a vector consisting of fading terms H_m associated with $\hat{c}_{m,i}$, and $\mathbf{z} \sim CN(0, \sigma_Z^2\mathbb{I})$, with identical matrix $\mathbb{I}$, is the noise vector. Depending on the data-rate modes, $\hat{\mathbf{c}}$, $\mathbf{h}$, and $\mathbf{z}$ are different and will be classified later. These modes share the same detection rule, the maximum likelihood detection, in which the symbol detected is

$$\tilde{c}_{m,i} = \underset{c_{m,i}}{\text{argmin}} \|\hat{\mathbf{c}} - c_{m,i}\mathbf{h}\|^2. \tag{7.64}$$

Since the system employs QPSK modulation, the average BER, denoted as P_b, is determined through the average symbol error rate P_s as $P_b = P_s$ [Bar04], where P_s is determined by averaging the symbol error rate given random vector $\mathbf{h}$ [i.e., $P_s = E\{P_s(\mathbf{h})\}$]. Based on the detection rule [Pro01], we have

$$P_s(\mathbf{h}) = \Pr\left[\text{Re}\{(c_{m,i} - \tilde{c}_{m,i})^*\mathbf{h}^H\mathbf{z}\} < -\frac{1}{2}|c_{m,i} - \tilde{c}_{m,i}|^2\|\mathbf{h}\|^2\right], \tag{7.65}$$

where $\text{Re}(x)$ yields the real component of the complex-valued x, $\|\cdot\|$ represents the Frobeniusnorm, and H represents a Hermitian. Because $\mathbf{z} \sim CN(0, \sigma_Z^2\mathbb{I})$, $\text{Re}\{(c_{m,i} - \tilde{c}_{m,i})^*\mathbf{h}^H\mathbf{z}\} \sim N(0, 1/2|c_{m,i} - \tilde{c}_{m,i}|^2\|\mathbf{h}\|^2\sigma_Z^2)$. After simplifications, we can show that

$$P_s(\mathbf{h}) = Q(\sqrt{2\rho}), \tag{7.66}$$

where $Q(\cdot)$ represents the Gaussian error function, defined in (3.60). In (7.66), ρ is defined as

$$\rho = \|\mathbf{h}\|^2 \frac{E_b}{\sigma_Z^2}, \qquad (7.67)$$

using the fact that the distance between QPSK symbols $|c_{m,i} - \tilde{c}_{m,i}|$ relates to the energy per bit E_b as $|c_{m,i} - \tilde{c}_{m,i}| = 2\sqrt{E_b}$. Our remaining task is to determine the probability density function (PDF) $f_\rho(t)$ of the random variable ρ for the three data-rate modes. The average symbol error rate then is given by [Bar04, Pro01]

$$P_s = \int_0^\infty f_\rho(t) Q(\sqrt{2t}) \, dt. \qquad (7.68)$$

Note that when ρ is a chi-square random variable with $2m$ degrees of freedom, its PDF is [Bar04]

$$f_\rho(t) = \frac{1}{(m-1)!\,(\overline{\gamma_\rho})^m} t^{m-1} e^{-t/\overline{\gamma_\rho}} \qquad \text{for } t \geq 0, \qquad (7.69)$$

where $\overline{\gamma_\rho} \triangleq \mathrm{E}\{\rho\}$ is the expectation of ρ corresponding to $m = 1$. Thus, the average symbol error rate is [Bar04]

$$P_s = p^m \sum_{k=0}^{m-1} \binom{m-1+k}{k} (1-p)^k, \qquad (7.70)$$

where

$$p \triangleq 1/2 \left(1 - \sqrt{\frac{\overline{\gamma_\rho}}{1 + \overline{\gamma_\rho}}} \right). \qquad (7.71)$$

In the following we demonstrate that ρ is approximately chi-square distributed with $2m$ degrees of freedom, where m is the overall spreading gain.

7.3.1 Overall Spreading Gain of 1

In this case, each frequency carrier and each time slot are used to transmit different information. The quantities in (7.63) are $\hat{\mathbf{c}} = \hat{c}_{m,i}$, $\mathbf{h} = H_m$, and $\mathbf{z} = z$. Thus, $\rho = E_b/\sigma^2_Z |H_m|^2$. From (7.14) we can rewrite the fading term as

$$H_m = \frac{1}{-j2\pi\varepsilon} w^{\varepsilon T_C} w^{-(m+\varepsilon)\tau} \mathbf{w}^H \mathbb{T} \, \mathbf{a}, \qquad (7.72)$$

where

$$\mathbf{w} = \left[w^{m(T_0+\tau_{0,0}+\tau)}, w^{m(T_0+\tau_{0,1}+\tau)}, \ldots, w^{m(T_L+\tau_{K,L}+\tau)} \right]^{\mathrm{T}} \qquad (7.73)$$

$$\mathbb{T} = \mathrm{diag}\left(e^{-j2\pi\varepsilon} w^{\varepsilon(T_0+\tau_{0,0}+\tau)} - 1, \ldots, e^{-j2\pi\varepsilon} w^{\varepsilon(T_{l_0}+\tau_{K,l_0}+\tau)} - 1, \right.$$

$$\left. e^{-j2\pi\varepsilon} w^{\varepsilon(T_{l_0+1}+\tau_{0,l_0+1}+\tau)} - 1, \ldots, e^{-j2\pi\varepsilon} w^{\varepsilon(T_L+\tau_{K,L}+\tau)} - 1 \right) \qquad (7.74)$$

$$\mathbf{a} = [\alpha_{0,0}, \alpha_{0,1}, \ldots, \alpha_{K,L}]^{\mathrm{T}}. \tag{7.75}$$

Because $\alpha_{k,l} \sim CN(0, {}_{k,l})$, where ${}_{k,l}$ follows (2.17), we have

$$\mathbf{a} = \mathbf{\Omega}^{1/2}\mathbf{b} \tag{7.76}$$

where

$$\mathbf{\Omega}^{1/2}\,\mathbf{\Omega}^{1/2} = \mathbf{\Omega} = \begin{pmatrix} \Omega_{0,0} & 0 & \cdots & 0 \\ 0 & \Omega_{0,1} & \cdots & 0 \\ \cdots & \cdots & \cdots & \cdots \\ 0 & 0 & \cdots & \Omega_{K,L} \end{pmatrix} \tag{7.77}$$

and

$$\mathbf{b} = [\beta'_{0,0},\ \beta'_{0,1}, \ldots, \beta'_{K,L}]^{\mathrm{T}}, \tag{7.78}$$

in which $\beta'_{k,l} \sim CN(0, 1)$. Therefore,

$$H_m = \frac{1}{-j2\pi\varepsilon}\, w^{\varepsilon Tc}\, w^{-(m+\varepsilon)\tau}\mathbf{w}^H\mathbb{T}\,\mathbf{\Omega}^{1/2}\mathbf{b}, \tag{7.79}$$

and consequently,

$$\rho = \frac{E_b}{\sigma_Z^2}\frac{1}{4\pi^2\varepsilon^2}\mathbf{b}^H\mathbf{\Omega}^{1/2}\mathbb{T}\mathbf{w}\mathbf{w}^H\mathbb{T}\mathbf{\Omega}^{1/2}\mathbf{b}. \tag{7.80}$$

Let us define $\mathbf{\Psi} = \mathbf{\Omega}^{1/2}\mathbb{T}\mathbf{w}\mathbf{w}^H\mathbb{T}\mathbf{\Omega}^{1/2}$. Clearly, $\mathbf{\Psi}$ is a nonnegative definite Hermitian matrix. Based on the singular-value decomposition theorem [Hor85], we can express $\mathbf{\Psi} = \mathbb{V}\mathbf{\Lambda}\mathbb{V}^H$, where $\mathbf{\Lambda}$ is a diagonal matrix containing the real and nonnegative eigenvalues of $\mathbf{\Psi}$ and $\mathbb{V}$ is a unitary matrix containing the eigenvectors associating with the eigenvalues in $\mathbf{\Lambda}$. Since rank$(\mathbf{\Psi}) \le \min\{\text{rank}(\mathbf{\Omega}^{1/2}), \text{rank}(\mathbb{T}), \text{rank}(\mathbf{w})\}$ where rank$(\mathbf{\Omega}^{1/2}) = \text{rank}(\mathbb{T}) = (K+1)(L+1)$ and rank$(\mathbf{w}) = 1$, there exists in $\mathbf{\Lambda}$ only one nonzero eigenvalue, which can be evaluated as

$$\text{eig}(\mathbf{\Psi}) = \sum_{l=0}^{l_0}\sum_{k=0}^{K}\Omega_{k,l}[2 - (e^{-j2\pi\varepsilon}w^{\varepsilon X_{k,l}} + e^{j2\pi\varepsilon}w^{-\varepsilon X_{k,l}})]$$

$$+ \sum_{l=l_0+1}^{L}\sum_{k=0}^{K}\Omega_{k,l}[2 - (e^{-j2\pi\varepsilon}w^{-\varepsilon X_{k,l}} + e^{j2\pi\varepsilon}w^{\varepsilon X_{k,l}})].$$

Thus, substituting ${}_{k,l}$ from (2.17) yields

$$\rho = \frac{E_b}{\sigma_Z^2}\frac{1}{4\pi^2\varepsilon^2}\left(\sum_{l=0}^{l_0}\sum_{k=0}^{K}\Omega_{0,0}e^{-T_l/\Gamma-\tau_{k,l}/\gamma}\,[2 - (e^{-j2\pi\varepsilon}w^{\varepsilon X_{k,l}} + e^{j2\pi\varepsilon}w^{-\varepsilon X_{k,l}})]\right.$$

$$\left.+ \sum_{l=l_0+1}^{L}\sum_{k=0}^{K}\Omega_{0,0}e^{-T_l/\Gamma-\tau_{k,l}/\gamma}[2 - (e^{-j2\pi\varepsilon}w^{-\varepsilon X_{k,l}} + e^{j2\pi\varepsilon}w^{\varepsilon X_{k,l}})])\right)|\beta|^2,$$

$$\tag{7.81}$$

where $\beta \sim CN(0, 1)$.

Equation (7.81) reveals that ρ is not a chi-square random variable with two degrees of freedom as in the case of a Rayleigh fading channel [Bar04, Pro01]. Here ρ is a product of a chi-square random variable $|\beta|^2$ and another random variable that is the sum of many combinations of the k- and l-Erlang random variables T_l and $\tau_{k,l}$. Hence, finding the PDF of ρ is difficult, if not impossible.

To obtain a closed-form formulation of the BER performance, we employ the approximation approach in [Sir06b] as follows. From (7.67), ρ has a quadratic form. Using a representation of quadratic form in [Mat92] and noting that $E\{\mathbf{h}\} = 0$, we get

$$\rho \approx \frac{E_b}{\sigma_Z^2} \sum_{s=1}^{S} \mathrm{eig}_s(\mathbf{\Phi})|\mu_s|^2, \tag{7.82}$$

where $\mu_s \sim CN(0, 1)$ and S is the rank of matrix $\mathbf{\Phi}$, defined as $\mathbf{\Phi} = E\{\mathbf{h}\mathbf{h}^H\}$. For the case of the gain factor of 1, $\mathbf{h} = H_m$; thus, $\mathbf{\Phi} = \sigma_H^2$, which is the variance of the fading term. Consequently,

$$\rho \approx \frac{E_b \sigma_H^2}{\sigma_Z^2}|\mu|^2. \tag{7.83}$$

Since $\mu \sim CN(0, 1)$, $|\mu|^2$ has a chi-square probability distribution with two degrees of freedom. Hence, ρ is approximately chi-square distributed with two degrees of freedom. Equation (7.83) also reveals that the expectation of ρ is $\overline{\gamma}_\rho = \overline{\gamma}_b(\varepsilon, \tau)$, the average SNR per bit.

Based on (7.70) with $m = 1$ and (7.71), the average symbol error rate and hence the average BER for this case is

$$P_b = P_s \approx 1/2 \left(1 - \sqrt{\frac{\overline{\gamma}_b(\varepsilon, \tau)}{1 + \overline{\gamma}_b(\varepsilon, \tau)}}\right). \tag{7.84}$$

7.3.2 Overall Spreading Gain of 2

In this case the same information is transmitted in two consecutive time slots. In such a case, (7.63) has $\hat{\mathbf{c}} = [\hat{c}_{m,i}\ \hat{c}_{m,i+1}]^T$, $\mathbf{h} = [H_m\ H_m]^T$, and $\mathbf{z} \sim CN(0, \sigma_Z^2 \mathbb{I}_2)$, assuming that the fading terms at the same subcarrier index m are iid. Following the same procedures as in Section 7.3.1, we can show that

$$\rho = \frac{E_b}{\sigma_Z^2} \frac{1}{4\pi^2 \varepsilon^2} \left(\sum_{l=0}^{l_0}\sum_{k=0}^{K} \Omega_{0,0} e^{-T_l/\Gamma - \tau_{k,l}/\gamma}\ [2 - (e^{-j2\pi\varepsilon}w^{\varepsilon X_{k,l}} + e^{j2\pi\varepsilon}w^{-\varepsilon X_{k,l}})]\right.$$

$$\left. + \sum_{l=l_0+1}^{L}\sum_{k=0}^{K} \Omega_{0,0} e^{-T_l/\Gamma - \tau_{k,l}/\gamma}[2 - (e^{-j2\pi\varepsilon}w^{-\varepsilon X_{k,l}} + e^{j2\pi\varepsilon}w^{\varepsilon X_{k,l}})]\right)$$

$$\times (|\beta_1|^2 + |\beta_2|^2), \tag{7.85}$$

where $\beta_i \sim CN(0, 1)$, $i = 1, 2$. The result in (7.85) reveals that in this case ρ relates to a chi-square random variable with four degrees of freedom.

Observe that the matrix

$$\boldsymbol{\Phi} = E\{\mathbf{hh}^H\} = \begin{pmatrix} \sigma_H^2 & 0 \\ 0 & \sigma_H^2 \end{pmatrix} \tag{7.86}$$

has two eigenvalues $\text{eig}_1(\boldsymbol{\Phi}) = \text{eig}_2(\boldsymbol{\Phi}) = \sigma_H^2$. Hence, similar to Section 7.3.1, ρ can be approximated as

$$\rho \approx \frac{E_b \sigma_H^2}{\sigma_Z^2}(|\mu_1|^2 + |\mu_2|^2), \tag{7.87}$$

where $\mu_i \sim CN(0, 1)$, $i = 1, 2$. Therefore, the average symbol error rate and hence the average BER for the case of $m = 2$ can be approximated as

$$P_b = P_s \approx p^2(3 - 2p), \tag{7.88}$$

where p is defined in (7.71) with $\overline{\gamma_\rho} = \overline{\gamma_b}(\varepsilon, \tau)$.

7.3.3 Overall Spreading Gain of 4

In this case, the same information is transmitted four times using two frequency carriers and two consecutive time slots. Accordingly, (7.63) has $\hat{\mathbf{c}} = \begin{bmatrix} \hat{c}_{m,i} & \hat{c}_{m,i+1} & \hat{c}^*_{N-m-1,i} & \hat{c}^*_{N-m-1,i+1} \end{bmatrix}^T$, $\mathbf{h} = \begin{bmatrix} H_m & H_m & H^*_{N-m-1} & H^*_{N-m-1} \end{bmatrix}^T$, and $\mathbf{z} \sim CN(0, \sigma_Z^2 \mathbb{I}_4)$. Following the same procedures as in Section 7.3.1, we have

$$\rho = \frac{E_b}{\sigma_Z^2} \frac{1}{4\pi^2\varepsilon^2} \left(\sum_{l=0}^{l_0} \sum_{k=0}^{K} \Omega_{0,0} e^{-T_l/\Gamma - \tau_{k,l}/\gamma} [2 - (e^{-j2\pi\varepsilon} w^{\varepsilon X_{k,l}} + e^{j2\pi\varepsilon} w^{-\varepsilon X_{k,l}})] \right.$$

$$\left. + \sum_{l=l_0+1}^{L} \sum_{k=0}^{K} \Omega_{0,0} e^{-T_l/\Gamma - \tau_{k,l}/\gamma} [2 - (e^{-j2\pi\varepsilon} w^{-\varepsilon X_{k,l}} + e^{j2\pi\varepsilon} w^{\varepsilon X_{k,l}})] \right)$$

$$\times (|\beta_1|^2 + |\beta_2|^2 + |\beta_3|^2 + |\beta_4|^2), \tag{7.89}$$

where $\beta_i \sim CN(0, 1)$, $i = 1, 2, 3, 4$, and the matrix

$$\boldsymbol{\Phi} \triangleq E\{\mathbf{hh}^H\}$$

$$= \begin{pmatrix} \sigma_H^2 & 0 & R_{(N-m-1),m} & R_{(N-m-1),m} \\ 0 & \sigma_H^2 & R_{(N-m-1),m} & R_{(N-m-1),m} \\ R^*_{(N-m-1),m} & R^*_{(N-m-1),m} & \sigma_H^2 & 0 \\ R^*_{(N-m-1),m} & R^*_{(N-m-1),m} & 0 & \sigma_H^2 \end{pmatrix},$$

where $R_{(N-m-1),m} = E\{H_m H_{N-m-1}\}$ is the complementary correlation between the fading terms at subcarrier m and its symmetric conjugate at subcarrier $N - m - 1$.

From H_m in (7.14), we can show that

$$R_{(N-m-1),m} = \frac{1}{4\pi^2 \varepsilon^2} w^{(N-1)\tau} \sum_{l=0}^{l_0} \sum_{k=0}^{K} E\left\{\Omega_{0,0} e^{-T_l/\Gamma' - \tau_{k,l}/\gamma'}\right.$$

$$\times \left[2 - \left(e^{-j2\pi\varepsilon} w^{\varepsilon X_{k,l}} + e^{j2\pi\varepsilon} w^{\varepsilon X_{k,l}}\right)\right]\right\}$$

$$+ \frac{1}{4\pi^2 \varepsilon^2} w^{(N-1)\tau} \sum_{l=l_0+1}^{L} \sum_{k=0}^{K} E\left\{\Omega_{0,0} e^{-T_l/\Gamma' - \tau_{k,l}/\gamma'}\right.$$

$$\times \left[2 - \left(e^{-j2\pi\varepsilon} w^{-\varepsilon X_{k,l}} + e^{j2\pi\varepsilon} w^{\varepsilon X_{k,l}}\right)\right]\right\}$$

$$\triangleq R(N-1), \tag{7.90}$$

where Γ' and γ' are defined such that

$$\frac{1}{\Gamma'} = \frac{1}{\Gamma} + \frac{j2\pi(N-1)}{T_S} \quad \text{and}$$

$$\frac{1}{\gamma'} = \frac{1}{\gamma} + \frac{j2\pi(N-1)}{T_S}.$$

By replacing Γ and γ in (7.32) with Γ' and γ', respectively, and multiplying the equation by $w^{(N-1)\tau}$, we obtain $R(N-1)$ in (7.90). Hence, $R(N-1)$ can be expressed in terms of $A_1, A_2, B_1,$ and B_2. Particularly, for various frequency and timing synchronization conditions, we have:

1. Perfect frequency and timing synchronization (i.e., $\varepsilon = 0$ and $\tau = 0$):

$$R(N-1) = w^{(N-1)\tau}\left(\Omega_{0,0} + \frac{1}{T_S^2} A_2(T_S, 0, \lambda, \gamma')\right.$$

$$\left. + \frac{1}{T_S^2} A_2(T_S, 0, \Lambda, \Gamma') + \frac{1}{T_S^2} A_1(T_S, 0, \Lambda, \Gamma', \lambda, \gamma')\right)$$

2. Imperfect timing synchronization (i.e. $\varepsilon = 0$ and $\tau \neq 0$):

$$R(N-1) = w^{(N-1)\tau}\left(\Omega_{0,0}\frac{(\tau + T_S)^2}{T_S^2} + \frac{1}{T_S^2} A_2(-T_S - \tau, 0, \lambda, \gamma')\right.$$

$$+ \frac{1}{T_S^2} A_2(-T_S - \tau, 0, \Lambda, \Gamma') - \frac{1}{T_S^2} A_2(-T_S - \tau, l_0, \Lambda, \Gamma')$$

$$+ \frac{1}{T_S^2} A_2(T_S - \tau, l_0, \Lambda, \Gamma') + \frac{1}{T_S^2} A_1(-T_S - \tau, 0, \Lambda, \Gamma', \lambda, \gamma')$$

$$\left. - \frac{1}{T_S^2} A_1(-T_S - \tau, l_0, \Lambda, \Gamma', \lambda, \gamma') + \frac{1}{T_S^2} A_1(T_S - \tau, l_0, \Lambda, \Gamma', \lambda, \gamma')\right)$$

3. Imperfect frequency synchronization (i.e., $\varepsilon \neq 0$ and $\tau = 0$):

$$R(N - 1) = \frac{1}{4\pi^2\varepsilon^2} w^{(N-1)\tau} [2\Omega_{0,0}(1 - \cos 2\pi\varepsilon) + B_2(0, 0, 0, \varepsilon, \varepsilon, 0, \lambda, \gamma')$$
$$+ B_2(0, 0, 0, \varepsilon, \varepsilon, 0, \Lambda, \Gamma') + B_1(0, 0, 0, \varepsilon, \varepsilon, 0, \Lambda, \Gamma', \lambda, \gamma')]$$

4. Imperfect frequency and timing synchronization (i.e., $\varepsilon \neq 0$ and $\tau \neq 0$):

$$R(N - 1) = \frac{1}{4\pi^2\varepsilon^2} w^{(N-1)\tau} \left[2\Omega_{0,0} \left(1 - \cos \left(\frac{2\pi\varepsilon\tau}{T_S} + 2\pi\varepsilon \right) \right) \right.$$
$$+ B_2(-\tau, 0, 0, \varepsilon, -\varepsilon, 0, \lambda, \gamma') + B_2(-\tau, 0, 0, \varepsilon, -\varepsilon, 0, \Lambda, \Gamma')$$
$$- B_2(-\tau, 0, 0, \varepsilon, -\varepsilon, l_0, \Lambda, \Gamma') + B_2(-\tau, 0, 0, \varepsilon, \varepsilon, l_0, \Lambda, \Gamma')$$
$$+ B_1(-\tau, 0, 0, \varepsilon, -\varepsilon, 0, \Lambda, \Gamma', \lambda, \gamma')$$
$$- B_1(-\tau, 0, 0, \varepsilon, -\varepsilon, l_0, \Lambda, \Gamma', \lambda, \gamma')$$
$$+ \left. B_1(-\tau, 0, 0, \varepsilon, \varepsilon, l_0, \Lambda, \Gamma', \lambda, \gamma') \right]$$

Let us define $r \triangleq R(N - 1)/\sigma_H^2$ as the normalized complementary correlation. Then the eigenvalues of matrix $\mathbf{\Phi}$ can be shown as $\text{eig}(\mathbf{\Phi}) = \sigma_H^2 [1 + |r|, \ 1, \ 1, \ 1 - |r|]^T$. Consequently,

$$\rho \approx \frac{E_b\sigma_H^2}{\sigma_Z^2}((1 + |r|)|\mu_1|^2 + |\mu_2|^2 + |\mu_3|^2 + (1 - |r|)|\mu_4|^2), \tag{7.91}$$

where $\mu_i \sim CN(0, 1)$, $i = 1, 2, 3, 4$. Based on (7.91), the average symbol error rate can be determined in the two specific cases as follows.

1. When $|r| \ll 1$, r can be ignored in (7.91). Hence, ρ is approximately a chi-square random variable with eight degrees of freedom. Thus, from (7.70) with $m = 4$, the average symbol error rate, and hence the average BER in this case, is

$$P_b = P_s \approx p^4 \sum_{k=0}^{3} \binom{3+k}{k}(1 - p)^k, \tag{7.92}$$

where p is defined in (7.71) with $\overline{\gamma_\rho} = \overline{\gamma_b}(\varepsilon, \tau)$.

2. When r cannot be ignored, ρ is no longer chi-square distributed because it is not a sum of iid random variables. This reflects the fact that the fading in different subchannels is highly correlated. To find the average symbol error rate, we use the alternative representation of the Q-function [Sim00],

$$Q(x) = \frac{1}{\pi} \int_0^{\pi/2} \exp\left(-\frac{x^2}{2 \sin^2 \theta} \right) d\theta \qquad \text{for } x \geq 0.$$

The average symbol error rate can then be expressed as

$$P_s = \frac{1}{\pi} \int_0^{\pi/2} M_\rho \left(-\frac{1}{\sin^2 \theta} \right) d\theta, \tag{7.93}$$

where $M_\rho(s) = E\{e^{s\rho}\}$ is the moment generating function of ρ. Because ρ is the sum of independent chi-square random variables, it can be shown that

$$M_\rho(s) \approx \frac{1}{(1 - s\overline{\gamma_\rho}(1 + |r|))(1 - s\overline{\gamma_\rho})^2(1 - s\overline{\gamma_\rho}(1 - |r|))}, \tag{7.94}$$

where $\overline{\gamma_\rho} = \overline{\gamma_b}(\varepsilon, \tau)$. Consequently, the average symbol error rate, and hence the average BER, is

$$P_b = P_s \approx \frac{1}{\pi} \int_0^{\pi/2} \frac{\sin^8 \theta}{(\sin^2 \theta + \overline{\gamma_\rho}(1 + |r|))(\sin^2 \theta + \overline{\gamma_\rho})^2(\sin^2 \theta + \overline{\gamma_\rho}(1 - |r|))} d\theta. \tag{7.95}$$

For multiband OFDM systems, (7.92) and (9.95) yield similar results since the normalized complementary correlation r is relatively small. For example, in the case of perfect frequency and timing synchronization, $r = 0.0987$, 0.0141, 0.0018, and 7.8343^{-004}, computed for CM1, CM2, CM3, and CM4, respectively. Figure 7.3 shows the average BER for the perfect synchronization case in CM1, CM2, CM3, and CM4. The average BER is plotted using (7.92) in the dotted-diamond curve (denoted as "Approximate") and (7.95) in the solid curve (denoted as "Exact"). Clearly, the approximated BER closely matches the exact BER.

We have derived the average BER completely for multiband OFDM systems. Depending on the data-rate modes: high rate, middle rate, or low rate, the average BER follows (7.84), (7.88), or (7.92) or (7.95) (depending on the value of the normalized complementary correlation r). The numerical and simulation results of the average BER performance are presented and analyzed in the next section.

7.4 PERFORMANCE BOUND

As presented in Chapter 4, the baseband of multiband OFDM systems contains two main blocks: a channel coding (including the bit interleaver) and an OFDM modulation. The performance analysis of the OFDM modulation is presented in Section 7.3, where the average BER is derived. In this section we consider the performance of entire multiband OFDM systems, including the convolutional encoder and bit interleaver at the transmitter and the Viterbi decoder and the bit de-interleaver at the receiver. As shown in Fig. 7.4, a transmitted binary sequence is input into the channel encoder. Redundancy is added to the sequence to improve the SNR for better

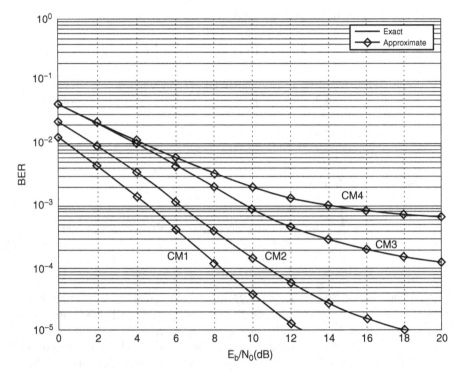

Figure 7.3 Average BER of multiband OFDM systems for the high-data-rate mode in channel model CM1.

detection at the receiver. For multiband OFDM systems, this process is done through convolutional coding and puncturing. The bit interleaver is used to prevent burst errors. After that, the binary sequence is modulated in the OFDM modulation and transmitted through the multipath fading channel with AWGN. At the receiver, after the OFDM demodulation, we obtain a binary sequence with a certain BER. This binary sequence is input into the bit de-interleaver to obtain an output sequence, which is then filled with dummy "zero" metrics in the de-puncturing process. After that, hard-decision Viterbi decoding is applied to obtain the binary sequence received.

The BER for the entire system can be bounded as [Pro01]

$$P_b^s \leq \frac{1}{M} \sum_{d=d_{\text{free}}}^{\infty} \beta_d P_d, \tag{7.96}$$

where M is the puncturing period, d_{free} is the free distance of the channel code, β_d is the weight spectrum representing the number of paths corresponding to the distance d, and P_d is the probability of selecting the incorrect path at distance d and can be

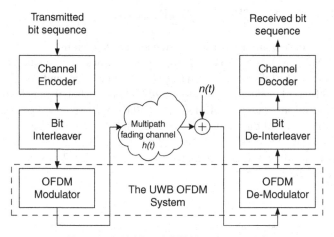

Figure 7.4 Multiband OFDM system model.

determined as

$$
P_d = \begin{cases}
\displaystyle\sum_{k=(d+1)/2}^{d} \binom{d}{k} P_b^k (1 - P_b)^{d-k} & d \text{ odd} \\[2em]
\displaystyle\frac{1}{2}\binom{d}{d/2} P_b^{d/2}(1 - P_b)^{d/2} + \sum_{k=d/2+1}^{d} \binom{d}{k} P_b^k (1 - P_b)^{d-k} & d \text{ even,}
\end{cases}
\tag{7.97}
$$

where

$$
P_b = \int_0^{\infty} f_\rho(t) Q(\sqrt{2t}) \, dt
\tag{7.98}
$$

is the average BER after the OFDM demodulation block, determined in Section 7.3. The remaining task is how to find the weight spectrum β_d corresponding to the convolutional code and the puncturing patterns in the system.

Theoretically, β_d can be found based on the transfer function of the convolutional code [Pro01]. The transfer function is obtained through solving the state equations. However, it is practically difficult to find the transfer function, especially for a convolutional code with large constraint length. For the mother code with coding rate $R = 1/3$ in our system, there are a total of 64 states. Thus, we need to derive and solve symbolically a linear system of 64 state equations for the transfer function. This is a difficult task. For the puncture codes in our system, obtaining the transfer functions is impossible. An alternative approach is to use computers to search for the free distance d_{free} and the weight spectrum β_d, for example, as being done in [Hag88]. If the information of d_{free} and β_d is available, following (7.96) would provide the performance bound of entire multiband OFDM systems.

7.5 NUMERICAL AND SIMULATION RESULTS

The performance of multiband OFDM systems is considered in UWB channel models: CM1, CM2, CM3, and CM4 with various conditions of frequency and timing synchronization. The OFDM system has $N = 128$ subcarriers and a subband bandwidth of 528 MHz. The durations of the useful OFDM symbol, the cyclic prefix, and the guard interval are $T_S = 242.42$ ns, $T_C = 60.61$ ns, and $T_G = 9.47$ ns, respectively. The total symbol duration is $T'_S = 312.5$ ns. The arrival rates Λ and λ and the decay factors Γ and γ of the cluster and ray, respectively, follow Table 2.1. The numerical results are presented and analyzed in Section 7.5.1. The simulation results are presented in Section 7.5.2.

7.5.1 Numerical Results

In Fig. 7.5 we plot the average BER versus SNR per bit E_b/N_0 of the UWB systems in the perfect frequency and timing synchronization in the four channel models. Two conclusions can be drawn from the figure. First, as the data rate increases, at the same SNR, the average BER also increases. This is due to the spreading gain that the data-rate mode inherits. The higher the spreading gain, the more the diversity order and hence the lower the average BER. The second conclusion from the figure is that the average BER increases as the severity of the channel increases. This is obvious in the figure. Among these channel models, CM1 is the least severe and has the lowest average BER, while CM4 is the most severe and results in the highest average BER compared at the same SNR.

In Fig. 7.6 we plot the average BER of UWB systems against the SNR per bit for the low-rate mode and various timing synchronization errors to illustrate UWB system performance in the imperfect timing synchronization. In the figure, $T = T_S/128$. Again, two conclusions can be drawn from the figure. First, positive timing errors always worsen the system performance, while small negative timing errors can improve it. As illustrated in Fig. 7.2(a), positive timing error corresponds to misplacement of the DFT window into the current OFDM symbol. Equivalently, timing synchronization is perfect, but the signal received arrives with an extra delay τ. Therefore, the current OFDM symbol receives more ISI from the previous OFDM symbol, and that worsens the system performance. On the other hand, a small negative timing error is equivalent to reduction of the channel delays and thus improves system performance. However, the increment in magnitude of negative timing error reduces the benefit. For large but negative timing error, the performance becomes worse again, as shown in Fig. 7.6. This is due to loss of the most powerful components, the first several rays in the first cluster in the arrival signal. This type of timing synchronization error increases the ISI. The second observation from the figure is that as the channel severity increases, larger negative timing errors are allowed. In Fig. 7.6(a), corresponding to CM1, the performance of $\tau = -6T$ is already worse than that of $\tau = 0T$, while the performance of $\tau = -9T$ corresponding to CM4 in Fig. 7.6(d) is still the best. This clear case is because CM4 has the largest channel delay,

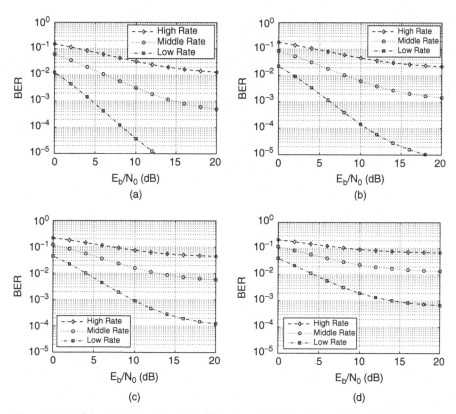

Figure 7.5 Average BER of multiband OFDM systems in perfect frequency and timing synchronization: (a) CM1; (b) CM2; (c) CM3; (d) CM4.

due to its extreme non-LOS condition, and its largest distance between the receiver and transmitter thus allows the largest negative timing error.

In Fig. 7.7 we plot the degradation ratio versus the relative carrier-frequency offset ε for the low-rate mode to illustrate system performance during imperfect frequency synchronization. The degradation ratio is plotted for the four channel models in SNRs of 0, 10, and 20 dB. Again, two conclusions can be drawn from the figure. First, as ε increases, the degradation ratio increases (i.e., the system performance becomes worse). The reason is that in terms of energy, the demodulated signal $\hat{c}_{m,i}$ contains fewer desired symbols $c_{m,i}$, due to its fading term H_m, while containing more undesired symbols, as illustrated in Fig. 7.2(b). As a result, frequency synchronization error increases the ICI and degrades the system performance. The second observation is that the degradation ratio decreases as the channel severity increases. In Fig. 7.7 the degradation ratio decreases as the SNR decreases. We see a similar trend across the channel models. The degradation ratio is the largest in CM1 and smallest in CM4. Since the performance is already very bad in the severe channel, the same amount of carrier-frequency offset causes a relatively small degradation in system performance.

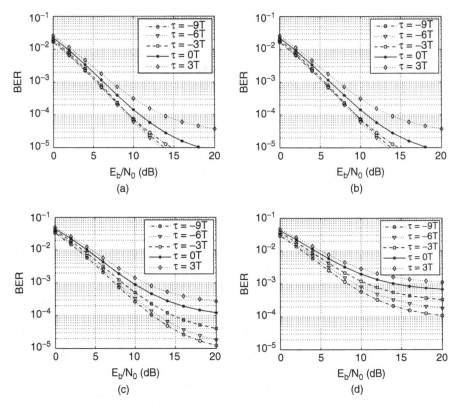

Figure 7.6 Average BER of multiband OFDM systems for the low-rate mode in imperfect timing synchronization: (a) CM1; (b) CM2; (c) CM3; (d) CM4.

The last figure of interest is Fig. 7.8, in which we plot the degradation ratio versus the relative carrier-frequency offset ε for the low-rate mode and various timing errors τ. The figure reveals that the frequency synchronization is more important than the timing synchronization. For instance, in Fig. 7.8(a), the degradation ratio in CM1 is 45 dB when $\varepsilon = \pm 1$, while it is about zero when $\tau = 3T$ (and $\varepsilon = 0$). Figure 7.2(b) shows that when ε is a nonzero integer, the demodulated signal totally loses its transmitted symbol. Thus, the degradation ratio should be infinite. However, in our cases, we are still able to receive the desired symbol. It is due to the channel multipath delay. It is shown in (7.14) that without the channel multipath delay (i.e., $T_l = 0$ and $\tau_{k,l} = 0$ for all l and k), and without a timing error (i.e., $\tau = 0$), the fading term H_m is identical to zero, and that leads to loss of the desired symbol.

7.5.2 Simulation and Numerical Results

The simulation results are plotted together with the numerical result in Fig. 7.9. The vertical axis is the average BER, and the horizontal axis is the SNR per bit, E_b/N_0.

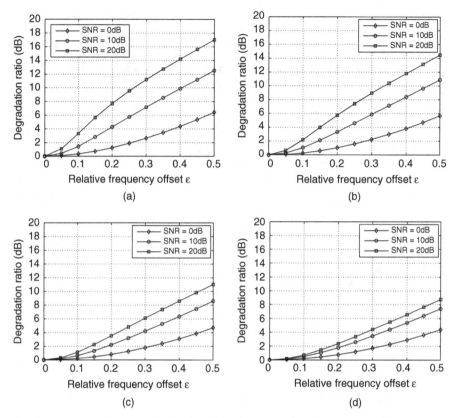

Figure 7.7 Degradation ratio of multiband OFDM systems for the low-rate mode in imperfect frequency synchronization: (a) CM1; (b) CM2; (c) CM3; (d) CM4.

So far we are able to obtain only the simulation average BER for UWB systems for high-rate mode in the case of channel model CM1 and perfect frequency and timing synchronization. The reason is that simulation consumes an enormous amount of time. As we have seen, our performance analysis is based on the continuous-time channel impulse response, which consists of a high average number of multipath delays. The numbers of delays for CM1, CM2, CM3, and CM4 are 295, 765, 1460, and 3930 on average, respectively. To observe the ISI effect, we need to take all multipath delays into account. The number of multipath delays together with the number of subcarriers ($N = 128$) requires a very large number of additions and multiplications for computation of the received signal $r(t)$ based on (7.3). In addition, computation of demodulated signal $\hat{c}_{m,i}$ requires integration of the received signal $r(t)$, as we see in (7.6). For the case in the figure, the integral is evaluated over about 10,000 subintervals for each $\hat{c}_{m,i}$. Nevertheless, Fig. 7.9 shows that the simulation results match the numerical result very well. The simulation validates our performance analysis.

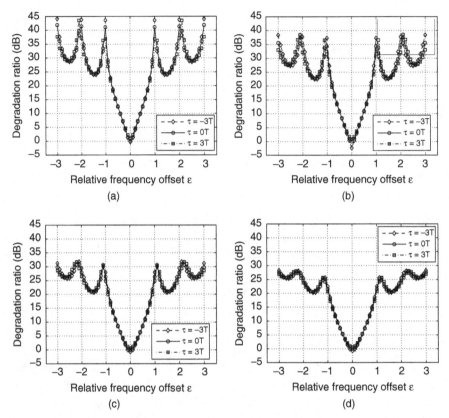

Figure 7.8 Degradation ratio of multiband OFDM systems for the low-rate mode in imperfect frequency and timing synchronization: (a) CM1; (b) CM2; (c) CM3; and (d) CM4.

7.6 CHAPTER SUMMARY

This chapter covers performance analysis of multiband OFDM systems in the four IEEE 802.15.3a channel models under four conditions of frequency and timing synchronization. We first derive the average SNR of the systems in the standard channel models. Then we analyze the system performance in terms of degradation ratio and average BER. The numerical results provide us a profound understanding of system performance in the standard channel models under different conditions of frequency and timing synchronization. In particular, the results show that small negative timing synchronization error can improve system performance, especially in the more severe channel models. In addition, frequency synchronization error is more sensitive since it degrades system performance more than timing synchronization error does. The simulation validates the theoretical analysis.

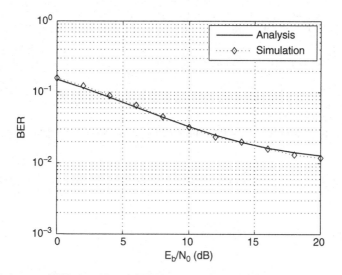

Figure 7.9 Average BER of multiband OFDM systems for the high-data-rate mode in channel model CM1.

APPENDIX: DERIVATIONS OF A_1, A_2, B_1, AND B_2

We define

$$A_1 \triangleq \sum_{l=l_0+1}^{\infty} \sum_{k=1}^{\infty} E\{\Omega_{0,0} e^{-T_l/\Gamma - \tau_{k,l}/\gamma} (T_l + \tau_{k,l} - T)^2\} \tag{7.A.1}$$

$$A_2 \triangleq \sum_{p=p_0+1}^{\infty} E\{\Omega_{0,0} e^{-X_p/\gamma_X} (X_p - T)^2\} \tag{7.A.2}$$

$$B_1 \triangleq \sum_{l=l_0+1}^{\infty} \sum_{k=1}^{\infty} E\left\{\Omega_{0,0} e^{-T_l/\Gamma - \tau_{k,l}/\gamma} \left[2 - \left(e^{-j2\pi\varepsilon'} e^{-j2\pi(n-m-\varepsilon)(T_l+\tau_{k,l}-T)/T_S}\right.\right.\right.$$
$$\left.\left.\left. + e^{j2\pi\varepsilon'} e^{\frac{j2\pi(n-m-\varepsilon)(T_l+\tau_{k,l}-T)}{T_S}}\right)\right]\right\} \tag{7.A.3}$$

$$B_2 \triangleq \sum_{p=p_0+1}^{\infty} E\left\{\Omega_{0,0} e^{-X_p/\gamma_X} \left[2 - \left(e^{-j2\pi\varepsilon'} e^{-j2\pi(n-m-\varepsilon)(X_p-T)/T_S}\right.\right.\right.$$
$$\left.\left.\left. + e^{j2\pi\varepsilon'} e^{\frac{j2\pi(n-m-\varepsilon)(X_p-T)}{T_S}}\right)\right]\right\}, \tag{7.A.4}$$

where $E\{\cdot\}$ denotes expectation, and T_l, $\tau_{k,l}$, and X_p are the arrival times in Poisson processes whose rates are Λ, λ, and λ_X and decay factors are Γ, γ, and γ_X, respectively.

Note that for a q-Erlang random variable Y_q with a parameter λ_Y [Leo94],

$$M_{Y_q}(s) = E\{e^{sY_q}\} = \frac{\lambda_Y^q}{(\lambda_Y - s)^q} \quad \text{for } q \geq 1 \tag{7.A.5}$$

is the moment generating function (MGF). Taking the nth derivative of the MGF, we have

$$E\left\{Y_q^n e^{sY_q}\right\} = \frac{d^{(n)} M_{Y_q}(s)}{ds} = \frac{(q+n-1)!}{(q-1)!} \frac{\lambda_Y^q}{(\lambda_Y - s)^{q+n}}. \tag{7.A.6}$$

Thus,

$$E\{Y_q e^{sY_q}\} = \frac{q\lambda_Y^q}{(\lambda_Y - s)^{q+1}} \tag{7.A.7}$$

and

$$E\left\{Y_q^2 e^{sY_q}\right\} = \frac{q(q+1)\lambda_Y^q}{(\lambda_Y - s)^{q+2}}. \tag{7.A.8}$$

We use (7.A.5), (7.A.7), and (7.A.8) in the following derivations.

A.1 Derivation of A_1 and A_2

Using the fact that T_l and $\tau_{k,l}$ are statistically independent, we rewrite

$$\begin{aligned}
A_1 = \Omega_{0,0} \sum_{l=l_0+1}^{\infty} \sum_{k=1}^{\infty} &\left[E\left\{T_l^2 e^{-T_l/\Gamma}\right\} E\left\{e^{-\tau_{k,l}/\gamma}\right\} + E\left\{e^{-T_l/\Gamma}\right\} E\left\{\tau_{k,l}^2 e^{-\tau_{k,l}/\gamma}\right\} \right.\\
&+ T^2 E\left\{e^{-T_l/\Gamma}\right\} E\left\{e^{-\tau_{k,l}/\gamma}\right\} + E\left\{T_l e^{-T_l/\Gamma}\right\} E\left\{\tau_{k,l} e^{-\tau_{k,l}/\gamma}\right\} \\
&\left. - 2T E\left\{T_l e^{-T_l/\Gamma}\right\} E\left\{e^{-\tau_{k,l}/\gamma}\right\} - 2T E\left\{e^{-T_l/\Gamma}\right\} E\left\{\tau_{k,l} e^{-\tau_{k,l}/\gamma}\right\} \right]. \tag{7.A.9}
\end{aligned}$$

Now, the results in (7.A.5), (7.A.7), and (7.A.8) allow us to have

$$\begin{aligned}
A_1 = \Omega_{0,0} \left[\right. & \sum_{l=l_0+1}^{\infty} \frac{l(l+1)\Lambda^l}{(\Lambda + 1/\Gamma)^{l+2}} \sum_{k=1}^{\infty} \frac{\lambda^k}{(\lambda + 1/\gamma)^k} \\
&+ \sum_{l=l_0+1}^{\infty} \frac{\Lambda^l}{(\Lambda + 1/\Gamma)^l} \sum_{k=1}^{\infty} \frac{k(k+1)\lambda^k}{(\lambda + 1/\gamma)^{k+2}} \\
&+ T_S^2 \sum_{l=l_0+1}^{\infty} \frac{\Lambda^l}{(\Lambda + 1/\Gamma)^l} \sum_{k=1}^{\infty} \frac{\lambda^k}{(\lambda + 1/\gamma)^k}
\end{aligned}$$

$$+2 \sum_{l=l_0+1}^{\infty} \frac{l\Lambda^l}{(\Lambda + 1/\Gamma)^{l+1}} \sum_{k=1}^{\infty} \frac{k\lambda^k}{(\lambda + 1/\gamma)^{k+1}}$$

$$-2T \sum_{l=l_0+1}^{\infty} \frac{l\Lambda^l}{(\Lambda + 1/\Gamma)^{l+1}} \sum_{k=1}^{K} \frac{\lambda^k}{(\lambda + 1/\gamma)^k}$$

$$-2T \left. \sum_{l=l_0+1}^{\infty} \frac{\Lambda^l}{(\Lambda + 1/\Gamma)^l} \sum_{k=1}^{\infty} \frac{k\lambda^k}{(\lambda + 1/\gamma)^{k+1}} \right]. \tag{7.A.10}$$

Our next step is to evaluate the geometric series. Note that for $\beta = \lambda_Y \gamma_Y / (\lambda_Y \gamma_Y + 1)$ ($|\beta| < 1$),

$$\sum_{q=0}^{\infty} \beta^q = \frac{1}{1-\beta} = (\lambda_Y \gamma_Y + 1)$$

$$\sum_{q=0}^{\infty} q\beta^q = \frac{\beta}{(1-\beta)^2} = \lambda_Y \gamma_Y (\lambda_Y \gamma_Y + 1)$$

$$\sum_{q=0}^{\infty} q(q+1)\beta^q = \frac{2\beta}{(1-\beta)^3} = 2\lambda_Y \gamma_Y (\lambda_Y \gamma_Y + 1)^2.$$

Hence,

$$\sum_{q=q_0+1}^{\infty} \beta^q = \beta^{q_0+1} \sum_{p=0}^{\infty} \beta^q$$

$$= f_1(q_0, \lambda_Y, \gamma_Y) \tag{7.A.11}$$

$$\sum_{q=q_0+1}^{\infty} q\beta^q = \beta^{q_0+1} \left[\sum_{p=0}^{\infty} q\beta^q + (q_0 + 1) \sum_{p=0}^{\infty} \beta^q \right]$$

$$= (\lambda_Y \gamma_Y + 1) f_2(q_0, \lambda_Y, \gamma_Y) \tag{7.A.12}$$

$$\sum_{q=q_0+1}^{\infty} q(q+1)\beta^q = \beta^{q_0+1} \left[\sum_{p=0}^{\infty} q(q+1)\beta^q + 2(q_0 + 1) \sum_{p=0}^{\infty} q\beta^q \right.$$

$$\left. + (q_0 + 1)(q_0 + 2) \sum_{p=0}^{\infty} \beta^q \right]$$

$$= (\lambda_Y \gamma_Y + 1)^2 f_3(q_0, \lambda_Y, \gamma_Y), \tag{7.A.13}$$

where we define

$$f_1(q_0, \lambda_Y, \gamma_Y) \triangleq \frac{(\lambda_Y \gamma_Y)^{q_0+1}}{(\lambda_Y \gamma_Y + 1)^{q_0+1}} (\lambda_Y \gamma_Y + 1) \tag{7.A.14}$$

$$f_2(q_0, \lambda_Y, \gamma_Y) \triangleq \frac{(\lambda_Y \gamma_Y)^{q_0+1}}{(\lambda_Y \gamma_Y + 1)^{q_0+1}} (\lambda_Y \gamma_Y + q_0 + 1) \tag{7.A.15}$$

$$f_3(q_0, \lambda_Y, \gamma_Y) \triangleq \frac{(\lambda_Y \gamma_Y)^{q_0+1}}{(\lambda_Y \gamma_Y + 1)^{q_0+2}} [2\lambda_Y \gamma_Y (\lambda_Y \gamma_Y + 1)$$
$$+ 2(q_0 + 1)\lambda_Y \gamma_Y + (q_0 + 1)(q_0 + 2)]. \tag{7.A.16}$$

Applying the results in (7.A.11), (7.A.12), and (7.A.13) to (7.A.10), we finally have

$$A_1 = \Omega_{0,0} \big[\Gamma^2 f_3(l_0, \Lambda, \Gamma) f_1(0, \lambda, \gamma) + \gamma^2 f_1(l_0, \Lambda, \Gamma) f_3(0, \lambda, \gamma)$$
$$+ T^2 f_1(l_0, \Lambda, \Gamma) f_1(0, \lambda, \gamma) + 2\Gamma\gamma f_2(l_0, \Lambda, \Gamma) f_2(0, \lambda, \gamma)$$
$$- 2T\Gamma f_2(l_0, \Lambda, \Gamma) f_1(0, \lambda, \gamma) - 2T\gamma f_1(l_0, \Lambda, \Gamma) f_2(0, \lambda, \gamma) \big]. \tag{7.A.17}$$

In a similar manner, we are able to show that

$$A_2 = \Omega_{0,0} \big[\gamma_X^2 f_3(p_0, \lambda_X, \gamma_X) + T^2 f_1(p_0, \lambda_X, \gamma_X) - 2T\gamma_X f_2(p_0, \lambda_X, \gamma_X) \big]. \tag{7.A.18}$$

A.2 Derivation of B_1 and B_2

The derivation of B_1 and B_2 follows the same procedure as that of the A_1 and A_2. Since T_l and $\tau_{k,l}$ are statistically independent,

$$B_1 = \Omega_{0,0} \Bigg[2 \sum_{l=l_0+1}^{\infty} E\{e^{-T_l/\Gamma}\} \sum_{k=1}^{\infty} E\{e^{-\tau_{k,l}/\gamma}\}$$

$$- e^{-j2\pi\varepsilon'} e^{\frac{j2\pi(n-m-\varepsilon)T}{T_S}} \sum_{l=l_0+1}^{\infty} E\left\{ e^{(-1/\Gamma - j2\pi(n-m-\varepsilon)/T_S)T_l} \right\}$$

$$\times \sum_{k=1}^{\infty} E\left\{ e^{(-1/\gamma - j2\pi(n-m-\varepsilon)/T_S)\tau_{k,l}} \right\} - e^{-j2\pi\varepsilon'} e^{j2\pi(n-m-\varepsilon)T/T_S}$$

$$\times \sum_{l=l_0+1}^{\infty} E\left\{ e^{(-1/\Gamma + j2\pi(n-m-\varepsilon)/T_S)T_l} \right\} \sum_{k=1}^{\infty} E\left\{ e^{(-1/\gamma + j2\pi(n-m-\varepsilon)/T_S)\tau_{k,l}} \right\} \Bigg]. \tag{7.A.19}$$

For a q-Erlang random variable Y_q, using (7.A.5) we have

$$E\left\{e^{(-1/\gamma_Y \pm j2\pi(n-m-\varepsilon)/T_S)Y_q}\right\} = \frac{\lambda_Y^q}{(\lambda_Y + 1/\gamma_Y \mp j2\pi(n-m-\varepsilon)/T_S)^q}$$

$$= \left(\beta_Y e^{\pm j\theta_Y}\right)^q , \tag{7.A.20}$$

where we define

$$\beta_Y \triangleq \frac{\lambda_Y}{\sqrt{(\lambda_Y + 1/\gamma_Y)^2 + 4\pi^2(n-m-\varepsilon)^2/T_S^2}} \tag{7.A.21}$$

$$\theta_Y \triangleq \arctan\left(\frac{2\pi(n-m-\varepsilon)}{T_S}\frac{\gamma_Y}{\lambda_Y\gamma_Y + 1}\right). \tag{7.A.22}$$

Therefore, we are able to show that

$$\sum_{q=0}^{\infty} \left(\beta_Y e^{\pm j\theta_Y}\right)^q = \frac{1 - \beta_Y e^{\mp j\theta_Y}}{1 + \beta_Y^2 - 2\beta_Y \cos\theta_Y}. \tag{7.A.23}$$

Combining (7.A.11), (7.A.20), and (7.A.23), we have

$$\sum_{q=q_0+1}^{\infty} E\left\{e^{(-1/\gamma_Y \pm j2\pi(n-m-\varepsilon)/T_S)Y_q}\right\} = \beta_Y^{q_0+1} e^{\pm j(q_0+1)\theta_Y} \frac{1 - \beta_Y e^{\mp j\theta_Y}}{1 + \beta_Y^2 - 2\beta_Y \cos\theta_Y}. \tag{7.A.24}$$

Using (7.A.5) and (7.A.24) in (7.A.19) and after some simplifications, we obtain

$$B_1 = 2\Omega_{0,0} f_1(l_0, \Lambda, \Gamma) f_1(0, \lambda, \lambda) - 2\Omega_{0,0} \beta_T^{l_0+1} \beta_\tau$$

$$\times \left[\frac{\cos((l_0+1)\theta_T + \theta_\tau - (2\pi(n-m-\varepsilon)T/T_S) + 2\pi\varepsilon')}{\left(1 + \beta_T^2 - 2\beta_T\cos\theta_T\right)\left(1 + \beta_\tau^2 - 2\beta_\tau\cos\theta_\tau\right)} \right.$$

$$- \frac{\beta_\tau \cos((l_0+1)\theta_T - (2\pi(n-m-\varepsilon)T/T_S) + 2\pi\varepsilon')}{\left(1 + \beta_T^2 - 2\beta_T\cos\theta_T\right)\left(1 + \beta_\tau^2 - 2\beta_\tau\cos\theta_\tau\right)}$$

$$- \frac{\beta_T \cos(l_0\theta_T + \theta_\tau - (2\pi(n-m-\varepsilon)T/T_S) + 2\pi\varepsilon')}{\left(1 + \beta_T^2 - 2\beta_T\cos\theta_T\right)\left(1 + \beta_\tau^2 - 2\beta_\tau\cos\theta_\tau\right)}$$

$$\left. + \frac{\beta_T\beta_\tau \cos(l_0\theta_T - (2\pi(n-m-\varepsilon)T/T_S) + 2\pi\varepsilon')}{\left(1 + \beta_T^2 - 2\beta_T\cos\theta_T\right)\left(1 + \beta_\tau^2 - 2\beta_\tau\cos\theta_\tau\right)} \right]. \tag{7.A.25}$$

The derivation of B_2 is similar to that of B_1. The result is presented below.

$$B_2 = 2\Omega_{0,0} f_1(p_0, \lambda_X, \gamma_X) - 2\Omega_{0,0} \beta_X^{p_0+1}$$
$$\times \left[\frac{\cos((p_0+1)\theta_X - (2\pi(n-m-\varepsilon)T/T_S) + 2\pi\varepsilon')}{1 + \beta_X^2 - 2\beta_X \cos\theta_X} \right.$$
$$\left. -\beta_X \frac{\cos(p_0\theta_X - (2\pi(n-m-\varepsilon)T/T_S) + 2\pi\varepsilon')}{1 + \beta_X^2 - 2\beta_X \cos\theta_X} \right], \qquad (7.A.26)$$

where $f_1(\cdot)$, β_X, and θ_X follow (7.A.14), (7.A.21), and (7.A.22), respectively.

8

DIFFERENTIAL MULTIBAND OFDM

Since many applications enabled by UWB are expected to be in portable devices, low complexity becomes a fundamental requirement. This indicates the pressing need for a simple transceiver design. A conventional coherent detection system requires channel estimation and hence introduces complexity to the receiver. Moreover, even though channel estimates may be available when the channel changes slowly compared with the symbol rate, it may not be possible to acquire them in a fast-fading environment. An alternative approach to overcome such problems is through the use of noncoherent detection techniques. In recent years, noncoherent UWB systems have been proposed (e.g., in [Ho02]). Nevertheless, most of the existing work is based on single-band impulse radio technology. The current works for multiband approaches focus primarily on coherent detection schemes [Sir06a].

Differential modulation [Pro01] has been widely known as one of many practical alternatives that bypasses channel estimation in communications systems. MIMO-OFDM, on the other hand, has been shown to be an effective way to increase the capacity of a frequency-selective channel without sacrificing bandwidth. Recently, a technique of incorporating differential modulation with the MIMO-OFDM scheme, the *differential space–time–frequency* (DSTF) *scheme* [Dig02, Li03, Ma03, Hoc99, Wan02, Him05a, Him06, Su04b], was introduced as an effective, yet practical modulation scheme for frequency-selective channels. The DSTF modulation encodes simultaneously across spatial, temporal, and frequency domains such that both spatial and frequency diversity can be explored.

In this chapter we present a differential encoding and decoding scheme for UWB multiband OFDM systems referred to as DMB-OFDM [Him05b]. The scheme is described for a general multiple-antenna system with M_t transmitting and M_r receiving antennas. This includes the DMB-OFDM scheme for a single-antenna system as a special case with $M_t = 1$ and $M_r = 1$. In the DMB-OFDM scheme, the information is encoded jointly across spatial, temporal, and frequency domains. By differentially en/decoding in the frequency domain, the scheme does not rely on the assumption that the fading channel stays constant within several OFDM symbol durations. In contrast

Ultra-Wideband Communications Systems: Multiband OFDM Approach, By W. Pam Siriwongpairat and K. J. Ray Liu
Copyright © 2008 John Wiley & Sons, Inc.

Adapted with permission from © 2005 *IEEE Global Telecommunications Conference*, Vol. 6, Dec. 2005.

to conventional systems, the scheme proposed incorporates differential modulation with hopping multiband OFDM transmission. In this way, the scheme proposed explores not only spatial and frequency diversities, but also additional diversity from time-domain spreading. Hence, the scheme efficiently explores all the available spatial and frequency diversities richly inherent in UWB channels. Furthermore, we analyze the performance of the scheme in realistic UWB channels. Using the S-V channel model, we provide error probability characterization of the scheme that is able to capture the unique multipath-rich and random-clustering properties of UWB channels.

An overview of differential modulations is presented in Section 8.1. We begin with the differential modulation for single-antenna systems, followed by that for MIMO systems. In Section 8.2 we present a differential encoding and decoding scheme for multiband OFDM systems. The pairwise error probability is based on the S-V fading model in Section 8.3. We include some simulation results and discuss them in Section 8.4. Finally, in Section 8.5 we summarize the chapter.

8.1 DIFFERENTIAL MODULATION

8.1.1 Single-Antenna Systems

In single-antenna systems, noncoherent modulation is useful when knowledge of CSI is not available. The noncoherent modulation simplifies the receiver structure by omitting channel estimation and carrier or phase tracking. Some examples of the noncoherent modulation techniques are noncoherent frequency shift keying (NFSK) and differential modulation [Pro01]. Among these modulation techniques, the differential scheme is preferred to the SFSK because it provides better performance at the same operating SNR. Two classes of differential modulation schemes are available: differential M-ary quadrature amplitude modulation (DMQAM) and differential M-ary phase shift keying (DMPSK). In the DMQAM scheme, information is modulated through the amplitude difference among two consecutive symbols. The DMPSK, however, modulates information through the phase difference between two consecutive symbols. For a data rate of R bits per channel use, DMPSK signal constellation contains $Q = 2^R$ symbols. Each symbol v_q, $q \in \{0, Q - 1\}$, is generated by the gth root of unity: $v_q = e^{j2\pi q/Q}$.

The merit of the differential detection can be described as follows. Define s_τ as the differentially encoded symbol to be transmitted at time τ, α_τ as the fading coefficient, and n_τ as additive noise; then the signal received for the DMPSK modulation system can be written as

$$y_\tau = \sqrt{\rho}\, h_\tau s_\tau + w_\tau, \tag{8.1}$$

where $s_\tau = v_q s_{\tau-1}$ results from the differentially encoded information symbol v_q with the symbol $s_{\tau-1}$ transmitted previously, and $s_\tau = 1$ when $\tau = 0$. The DMPSK demodulator uses two consecutive received signals, $y_{\tau-1}$ and y_τ, to decode the information transmitted. Specifically, the DMPSK demodulator computes

$$y_\tau = \sqrt{\rho}\, h_\tau s_\tau + w_\tau = \sqrt{\rho}\, h_\tau v_q s_{\tau-1} + w_\tau, \tag{8.2}$$

where the term $s_{\tau-1}$ in the final equality of (8.2) can be reexpressed as

$$\sqrt{\rho}\, h_{\tau-1} s_{\tau-1} = y_{\tau-1} - w_{\tau-1}. \tag{8.3}$$

The efficient DMPSK demodulator relies on the assumption that the channel is almost constant over two symbol periods, $h_{\tau-1} \approx h_\tau$. Therefore, substituting (8.3) into (8.1), we have

$$\begin{aligned} y_\tau &= y_{\tau-1} v_q - v_q w_{\tau-1} + w_\tau \\ &= y_{\tau-1} v_q + \sqrt{2}\, \tilde{w}_\tau, \end{aligned} \tag{8.4}$$

where $\tilde{w}_\tau \triangleq (1/\sqrt{2})(w_\tau - v_q w_{\tau-1})$ is a complex Gaussian random variable with zero mean and unit variance. From (8.4), by treating $y_{\tau-1}$ as known channel information and v_q as a transmitted symbol, the expression in (8.4) can be considered as a received signal for coherent detection. In this way the optimum detector is simply the minimum distance detector by which to find a satisfactory symbol. Therefore, the differential decoder estimates the information transmitted using the following decoding rule:

$$v_{\hat{q}} = \operatorname*{argmin}_{q \in 0,1,\dots,Q-1} |y_\tau - v_q y_{\tau-1}|^2. \tag{8.5}$$

The sufficient statistic in (8.4) also provides an intuitive idea of the performance of the differential detection in comparison to its coherent counterpart. Specifically, the noise power is twice that of its coherent counterpart, so the SNR obtained is half of that with coherent detection. Therefore, the performance loss of 3 dB can be expected when using differential detection in comparison to the performance of its coherent counterpart. Nevertheless, the good trade-off in receiver complexity and performance has driven the differential scheme to be deployed in the IEEE IS-54 standard [EIA92] for cellular systems.

8.1.2 MIMO Systems

Recently, the merit of bypassing multichannel estimation attracted many researchers to naturally extend the single-antenna differential scheme for MIMO systems. For narrowband systems, research work on MIMO coding that does not require CSI at either the transmitter or the receiver is proposed in [Hoc00] as a unitary space–time modulation. It was shown in [Hoc00] that the unitary space–time modulation achieves the same diversity order as that of general space–time codings. In addition, the unitary signaling concept [Hoc00] has been generalized to a differential modulation for MIMO systems termed *differential unitary space–time* (DUST) *modulation* [Hoc01, Hug00]. The DUST scheme, which utilizes unitary group constellation, is suitable for MIMO systems with an arbitrary number of transmitting antennas.

The other class of differential MIMO systems is based on differential orthogonal space–time block codes (DSTBC) have been proposed in [Tar00] for two transmitting antennas and in [Jaf01] for more than two transmitting antennas. A special case of the works proposed in [Tar00] and [Jaf01] has been reported in [Gan02]. The related works in [Che03, Tao01] utilized multilevel amplitude modulation for DSTBC to improve the MIMO link performance. For wideband systems, the idea of employing

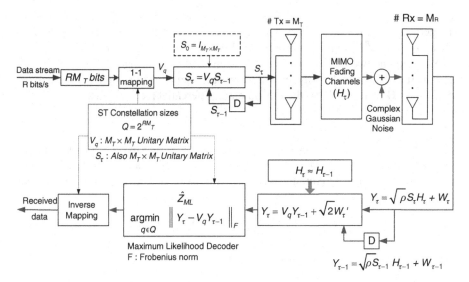

Figure 8.1 Descriptions of the differential unitary space–time modulation scheme.

DUST or DSTBC modulations with OFDM has been introduced in the form of differential space–time–frequency (DSTF) MIMO-OFDM systems [Dig02, Li03, Ma03, Hoc99, Wan02]. In addition, a differential scheme for MIMO-OFDM that is more robust to fast fading have been proposed in [Him06] and [Su04b] for time- and frequency-domain differential modulations, respectively.

The conceptual diagram of the DUST scheme is depicted in Fig. 8.1. The figure shows a MIMO system with M_T transmitting antennas and M_R receiving antennas. The channel state information is assumed unknown to either the transmitter or the receiver. For T consecutive time intervals, the received signals are formulated in matrix form as

$$\mathbf{Y}_\tau = \sqrt{\rho}\,\mathbf{S}_\tau \mathbf{H}_\tau + \mathbf{W}_\tau, \quad \tau = 0, 1, \ldots, \tag{8.6}$$

where τ is the time index of block transmissions, $\mathbf{Y}_\tau$ is the $T \times M_R$ received signal matrix, and $\mathbf{S}_\tau$ is the $T \times M_T$ transmitted signal matrix. The $M_T \times M_R$ fading-coefficient matrix $\mathbf{H}_\tau$ and the $T \times M_R$ additive noise matrix $\mathbf{W}_\tau$ have complex Gaussian elements with $\mathcal{C}N(0, 1)$ distributed. The signal transmitted is normalized to have unit energy during one transmission period to ensure that ρ is the average SNR per receiver (i.e., $\mathrm{E}[\sum_{i=1}^{M_T} |s_t^i|^2] = 1$), where E represents the expectation operator.

In the following we assume that the signal matrices transmitted are square (i.e., $T = M_T$). The transmission process follows the fundamental differential transmitter equation [Hoc01],

$$\mathbf{S}_\tau = \begin{cases} \mathbf{V}_q \mathbf{S}_{\tau-1}, & \tau = 1, 2, \ldots, \\ \mathbf{I}_{M_T \times M_T}, & \tau = 0, \end{cases} \tag{8.7}$$

where $\mathbf{I}_{M_T \times M_T}$ is an $M_T \times M_T$ identity matrix; $q \in \{0, 1, \ldots, Q-1\}$ denotes an integer index of a distinct unitary matrix signal $\mathbf{V}_q$ drawn from a signal constellation $\mathcal{V}$ of size $Q = 2^{RM_T}$, with R representing the information rate in bit/s $\cdot$ Hz.

We combine two consecutive received signal matrices using (8.6) and (8.7) and assume that the channel coefficients are almost constant over two consecutive blocks (i.e., $\mathbf{H}_\tau \approx \mathbf{H}_{\tau-1}$). We obtain the fundamental differential receiver equation,

$$\mathbf{Y}_\tau = \mathbf{V}_q \mathbf{Y}_{\tau-1} + \sqrt{2}\,\mathbf{W}'_\tau, \tag{8.8}$$

where $\mathbf{W}'_\tau = (1/\sqrt{2})(\mathbf{W}_\tau - \mathbf{V}_q \mathbf{W}_{\tau-1})$ is an $M_T \times M_R$ additive independent noise matrix with $\mathcal{CN}(0, 1)$ distributed elements. Similar to (8.4) and (8.5), the differential decoder performs maximum likelihood decoding, and the decision rule can be expressed as [Hoc01]

$$\widehat{z}_\tau^{ML} = \underset{q \in \{0,1,\ldots,Q-1\}}{\operatorname{argmin}} \|\mathbf{Y}_\tau - \mathbf{V}_q \mathbf{Y}_{\tau-1}\|_F, \tag{8.9}$$

where $\|\cdot\|_F$ denotes the Frobenius norm[1] [Hor85].

8.2 DIFFERENTIAL SCHEME FOR MULTIBAND OFDM SYSTEMS

In this section we present the DMB-OFDM scheme, a frequency-domain differential scheme for multiband OFDM system. In addition, we exploit the additional diversity from band hopping inherently in multiband transmission by jointly encoding across K OFDM blocks and transmitted the K OFDM symbols on different subbands. In each OFDM block, we exploit subcarrier interleaving strategy as in [Su04b].

8.2.1 System Model

We consider a peer-to-peer multiband OFDM system equipped with M_t transmitting and M_r receiving antennas. Within each subband, OFDM modulation with N subcarriers is used at each transmitting antenna. The channel model is based on the S-V model presented in Chapter 2.

We denote $x_i^k(n)$ as a differentially encoded data symbol to be transmitted on the nth subcarrier at the ith transmitting antenna during the kth OFDM symbol period. At the receiver, after cyclic prefix removal and OFDM demodulating, the signal received at the nth subcarrier at the jth receiver antenna during the kth OFDM block is given by

$$y_j^k(n) = \sqrt{\rho} \sum_{i=1}^{M_t} x_i^k(n) H_{ij}^k(n) + w_j^k(n), \tag{8.10}$$

where ρ is the average SNR per receiver and

$$H_{ij}^k(n) = \sum_{c=0}^{C} \sum_{l=0}^{L} \alpha_{ij}^k(c, l) \exp[-\mathbf{j}2\pi n\, \Delta f(T_c + \tau_{c,l})] \tag{8.11}$$

---

[1]Let matrix $\mathbf{A} = (a_{ij})$; then the Frobenius norm of $\mathbf{A}$ is $\|\mathbf{A}\|_F = \operatorname{tr}(\mathbf{A}^H \mathbf{A}) = \operatorname{tr}(\mathbf{A}\mathbf{A}^H) = \sum_{j=1}^{N} \sum_{i=1}^{M} |a_{ij}|^2$.

is the subchannel gain. Here $\Delta f = 1/T_s$ is the intersubcarrier spacing and T_s is the OFDM symbol period. The additive noise $w_j^k(n)$ is modeled as an independent complex Gaussian random variable with zero mean and unit variance.

8.2.2 Differential Encoding and Transmitting Signal Structure

We introduce a differential multiband OFDM scheme based on the transmitter signal structure proposed in [Sir06a]. Particularly, $\mathbf{X}$ is a jointly designed $KN \times M_t$ space–time–frequency code structure which consists of stacking space–frequency signal $\mathbf{X}^k$, each of dimension $N \times M_t$, for K OFDM symbols. To reduce the complexity of the design, we divide $\mathbf{X}^k$ into several submatrices or groups. By introducing a fixed integer G $(1 \le G \le N)$ as a number of jointly encodedsubcarriers, $\mathbf{X}^k$ at each OFDM symbol is partitioned into $P = \lfloor N/(GM_t) \rfloor$ submatrices as follows [Sir06a]:

$$\mathbf{X}^k = \left[\left(\mathbf{X}_1^k\right)^T \left(\mathbf{X}_2^k\right)^T \cdots \left(\mathbf{X}_P^k\right)^T \left(\mathbf{0}_{N-PGM_t}\right)^T \right]^T \tag{8.12}$$

for $k = 1, 2, \ldots, K$ and T denotes the matrix transposition. The $(N - PGM_t) \times M_t$ matrix $\mathbf{0}_{N-PGM_t}$ represents a zero-padding matrix to be inserted if N cannot be divided by GM_t. Each $GM_t \times M_t$ submatrix X_p^k for $k = 1, \ldots, K$ and $p = 1, 2, \ldots, P$ is modeled as

$$\mathbf{X}_p^k = \text{diag}\left(\mathbf{x}_{p,1}^k \mathbf{x}_{p,2}^k \cdots \mathbf{x}_{p,M_t}^k\right), \tag{8.13}$$

where $\text{diag}(\cdot)$ denotes a diagonal operation that places all vectors or scalar elements at the main diagonal matrix, and $x_{p,i}^k$ for $i = 1, 2, \ldots, M_t$ is a $G \times 1$ vector:

$$\mathbf{x}_{p,i}^k = \left[s_{p,(i-1)G+1}^k s_{p,(i-1)G+2}^k \cdots s_{p,iG}^k\right]^T, \tag{8.14}$$

in which all $s_{p,m}^k$, $m = 1, 2, \ldots, GM_t$, are differentially encoded symbols that will be specified later. We differentially encode across K OFDM symbols within each group, and our desired transmitter signal structure for the pth group after differential encoding is a $KGM_t \times M_t$ matrix:

$$\mathbf{X}_p = \left[\left(\mathbf{X}_p^1\right)^T \left(\mathbf{X}_p^2\right)^T \cdots \left(\mathbf{X}_p^K\right)^T\right]^T \tag{8.15}$$

in which the ith column contains encoded symbols to be transmitted at the ith transmitting antenna.

We now specify information matrices to be differentially encoded as follows. Let $\mathbf{V}_p$ denote a $KGM_t \times KGM_t$ unitary information matrix having diagonal form as

$$\mathbf{V}_p = \text{diag}([v_{p,1} v_{p,2} \cdots v_{p,KGM_t}]^T), \tag{8.16}$$

in which $v_{p,m}$ is an information symbol. We jointly design the data within each information matrix $\mathbf{V}_p$, but design the matrices $\mathbf{V}_p$'s independently for different p.

Let $\mathbf{S}_p$ be a $KGM_t \times KGM_t$ differentially encoded signal matrix. We recursively construct $\mathbf{S}_p$ by [Hoc01]

$$\mathbf{S}_p = \begin{cases} \mathbf{V}_p \mathbf{S}_{p-1}, & p \ge 1; \\ \mathbf{I}_{KGM_t}, & p = 0. \end{cases} \tag{8.17}$$

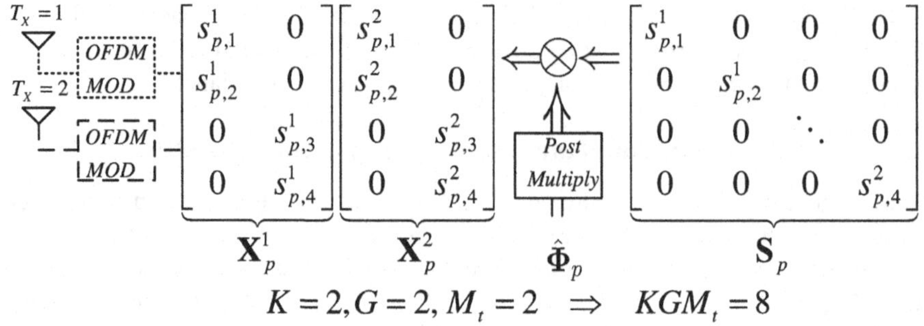

$$
\begin{array}{c}
T_X = 1 \\
\triangledown \quad \boxed{OFDM} \\
T_X = 2 \quad \boxed{MOD} \\
\triangledown \quad \dashv OFDM \vdash \\
\quad \quad \quad \lfloor MOD \rfloor
\end{array}
\begin{bmatrix}
s_{p,1}^1 & 0 \\
s_{p,2}^1 & 0 \\
0 & s_{p,3}^1 \\
0 & s_{p,4}^1
\end{bmatrix}
\begin{bmatrix}
s_{p,1}^2 & 0 \\
s_{p,2}^2 & 0 \\
0 & s_{p,3}^2 \\
0 & s_{p,4}^2
\end{bmatrix}
\Longleftarrow \otimes \Longleftarrow
\begin{array}{c} \\ \boxed{\begin{array}{c} Post \\ Multiply \end{array}} \\ \Uparrow \end{array}
\begin{bmatrix}
s_{p,1}^1 & 0 & 0 & 0 \\
0 & s_{p,2}^1 & 0 & 0 \\
0 & 0 & \ddots & 0 \\
0 & 0 & 0 & s_{p,4}^2
\end{bmatrix}
$$

$$
\mathbf{X}_p^1 \qquad \mathbf{X}_p^2 \qquad \hat{\mathbf{\Phi}}_p \qquad\qquad \mathbf{S}_p
$$

$$
K = 2, G = 2, M_t = 2 \implies KGM_t = 8
$$

Figure 8.2 Example of a differential encoded signal matrix and transmitter signal structure for a UWB multiband OFDM system with $K = 2$, $G = 2$, and $M_t = 2$.

Due to the diagonal structure of $\mathbf{V}_p$, $\mathbf{S}_p$ can be expressed as

$$
\mathbf{S}_p = \text{diag}\left(\left[s_{p,1}^1, \ldots, s_{p,GM_t}^1, \ldots, s_{p,1}^K, \ldots, s_{p,GM_t}^K\right]^{\mathrm{T}}\right), \tag{8.18}
$$

where $s_{p,m}^k$ is the differentially encoded complex symbol to be transmitted at subcarrier $(p - 1)GM_t + m$ during the kth OFDM block. To transform $\mathbf{S}_p$ into (8.15), we introduce a $KGM_t \times M_t$ multiplicative mapping matrix:

$$
\hat{\mathbf{\Phi}}_p = \mathbf{1}_K \otimes \mathbf{\Phi}_p, \tag{8.19}
$$

where $\mathbf{1}_K$ denotes a $K \times 1$ vector of all 1's, $\otimes$ denotes the Kronecker product [Hor85], $\mathbf{\Phi}_p = [\phi_1 \phi_2 \cdots \phi_{M_t}]$ is a $GM_t \times M_t$ mapping matrix in which $\phi_i = \mathbf{e}_i \otimes \mathbf{1}_G$ is a $GM_t \times 1$ vector and $\mathbf{e}_i$ is an $M_t \times 1$ unit vector whose ith component is 1 and all others are 0's. We postmultiply $\mathbf{S}_p$ by $\hat{\mathbf{\Phi}}_p$, resulting in the $KGM_t \times M_t$ matrix desired,

$$
\mathbf{X}_p = \mathbf{S}_p \hat{\mathbf{\Phi}}_p, \tag{8.20}
$$

as specified in (8.15). For better understanding if the concept of the DMB-OFDM scheme, we show in Fig. 8.2 an example of differentially encoded signals for $K = 2$, $G = 2$, and $M_t = 2$.

Before transmission, a subcarrier interleaving or subcarrier permutation scheme can be applied. As discussed in detail in [Liu03, Su05a], a subcarrier permutation scheme permutes the OFDM subcarrier symbols $\{x(1), x(2), \ldots, x(N - 1)\}$ such that the symbol $x(n)$ is transmitted in subcarrier $\sigma(n)$ instead of subcarrier n, where n, $\sigma(n) \in \{0, 1, \ldots, N - 1\}$. For example, using a subcarrier permutation scheme with a constant subcarrier spacing of 64 [Liu03, Su05a], the symbols in the subcarriers

$$
n = \{1, 2, 3, 4, 5, 6, \ldots 126, 127\} \tag{8.21}
$$

will be transmitted in the subcarriers

$$
\sigma(n) = \{1, 64, 2, 65, 3, 66, \ldots 63, 127\}. \tag{8.22}
$$

This allows the symbols encoded in a codeword to be transmitted over less correlated subcarriers. It also allows symbols that are successive in the direction of differential encoding to be transmitted over neighboring subcarriers that have a close channel frequency response.

8.2.3 Multiband Differential Decoding

The received signal vector corresponding to the transmitted matrix $\mathbf{X}_p$ is given by

$$\mathbf{y}_p = \sqrt{\rho}\left(\mathbf{I}_{M_r} \otimes \mathcal{D}(\mathbf{X}_p)\right)\mathbf{h}_p + \mathbf{w}_p, \tag{8.23}$$

where $\mathcal{D}(\mathbf{X_p})$ denotes an operation on an $KGM_t \times M_t$ matrix $\mathbf{X_p}$ that converts each column of $\mathbf{X_p}$ into a diagonal matrix and results in an $KGM_t \times KGM_tM_t$ matrix, expressed by

$$\mathcal{D}(\mathbf{X_p}) = \mathcal{D}([\mathbf{x}_{p,1} \cdots \mathbf{x}_{p,M_t}]) = [\text{diag}(\mathbf{x}_{p,1}) \cdots \text{diag}(\mathbf{x}_{p,M_t})]. \tag{8.24}$$

The matrix $\mathbf{h}_p = [(\mathbf{h}_{p,1})^T(\mathbf{h}_{p,2})^T \cdots (\mathbf{h}_{p,M_r})^T]^T$ is a channel matrix constructed from a $KGM_tM_t \times 1$ matrix:

$$\mathbf{h}_{p,j} = \left[\left(\mathbf{h}_{p,1j}^1\right)^T \cdots \left(\mathbf{h}_{p,1j}^K\right)^T \cdots \left(\mathbf{h}_{p,M_tj}^1\right)^T \cdots \left(\mathbf{h}_{p,M_tj}^K\right)^T\right], \tag{8.25}$$

where

$$\mathbf{h}_{p,ij}^k = \left[H_{ij}^k((p-1)GM_t) \cdots H_{ij}^k(pGM_t - 1)\right]^T \tag{8.26}$$

is a channel gain vector of size $GM_t \times 1$. The received signal matrix,

$$\mathbf{y}_p = \left[(\mathbf{y}_{p,1})^T(\mathbf{y}_{p,2})^T \cdots (\mathbf{y}_{p,M_r})^T\right]^T \tag{8.27}$$

is a $KGM_tM_r \times 1$ matrix constructed from the $KGM_t \times 1$ received signal vector

$$\mathbf{y}_{p,j} = \left[\left(\mathbf{y}_{p,j}^1\right)^T\left(\mathbf{y}_{p,j}^2\right)^T \cdots \left(\mathbf{y}_{p,j}^K\right)^T\right]^T, \tag{8.28}$$

in which $\mathbf{y}_{p,j}^k = [y_j^k((p-1)GM_t) \cdots y_j^k(pGM_t - 1)]^T$ is a $GM_t \times 1$ matrix. The noise matrix $\mathbf{w}_p$ is in the same form as $\mathbf{y}_p$, with $\mathbf{y}_{p,j}$ and $\mathbf{y}_{p,j}^k$ replaced by $\mathbf{w}_{p,j}$ and $\mathbf{w}_{p,j}^k$, respectively.

By substituting (8.20) into (8.23), we can reformulate $\mathbf{y}_p$ as

$$\mathbf{y}_p = \sqrt{\rho}\left(\mathbf{I}_{M_r} \otimes \mathcal{D}(\mathbf{S}_p\hat{\mathbf{\Phi}}_p)\right)\mathbf{h}_p + \mathbf{w}_p. \tag{8.29}$$

To simplify (8.29), we first observe from (8.19) that $\hat{\mathbf{\Phi}}_p$ can be reexpressed as $\hat{\mathbf{\Phi}}_p = [\tilde{\phi}_1\tilde{\phi}_2 \cdots \tilde{\phi}_{M_t}]$, where $\tilde{\phi}_i = \mathbf{1}_K \otimes \phi_i$. Therefore, $\mathcal{D}(\mathbf{S}_p\hat{\mathbf{\Phi}}_p)$ can be given by

$$\mathcal{D}(\mathbf{S}_p\hat{\mathbf{\Phi}}_p) = [\text{diag}(\mathbf{S}_p\tilde{\phi}_1) \cdots \text{diag}(\mathbf{S}_p\tilde{\phi}_{M_t})]. \tag{8.30}$$

According to (8.25) and (8.30) for each j, we have $\mathcal{D}(\mathbf{S}_p\hat{\mathbf{\Phi}}_p)\mathbf{h}_{p,j} = \sum_{i=1}^{M_t} \text{diag}(\mathbf{S}_p\tilde{\phi}_i)\mathbf{h}_{p,ij}$, which can be simplified to

$$\mathcal{D}(\mathbf{S}_p\hat{\mathbf{\Phi}}_p)\mathbf{h}_{p,j} = \mathbf{S}_p\sum_{i=1}^{M_t} \tilde{\phi}_i \circ \mathbf{h}_{p,ij} \triangleq \mathbf{S}_p\tilde{\mathbf{h}}_{p,j}, \tag{8.31}$$

where the last term on the right-hand side results from using the property of the Hadamardproduct [Hor85]. The $KG \times 1$ channel matrix $\tilde{\mathbf{h}}_{p,j}$ can be obtained by substituting (8.25) into (8.31) as

$$\tilde{\mathbf{h}}_{p,j} = \left[\left(\tilde{\mathbf{h}}_{p,1j}^1\right)^{\mathrm{T}} \cdots \left(\tilde{\mathbf{h}}_{p,1j}^K\right)^{\mathrm{T}} \cdots \left(\tilde{\mathbf{h}}_{p,M_tj}^1\right)^{\mathrm{T}} \cdots \left(\tilde{\mathbf{h}}_{p,M_tj}^K\right)^{\mathrm{T}} \right]^{\mathrm{T}}, \quad (8.32)$$

where

$$\tilde{\mathbf{h}}_{p,ij}^k = \left[H_{ij}^k\left(n_{p,i}^0\right) H_{ij}^k\left(n_{p,i}^1\right) \cdots H_{ij}^k\left(n_{p,i}^{G-1}\right) \right]^{\mathrm{T}} \quad (8.33)$$

is of size $G \times 1$, and $n_{p,i}^g = (i-1)G + (p-1)GM_t + g$ for $g = 0, 1, \ldots, G-1$. By denoting a $KGM_tM_r \times 1$ channel gain vector

$$\tilde{\mathbf{h}}_p = \left[(\tilde{\mathbf{h}}_{p,1})^{\mathrm{T}} (\tilde{\mathbf{h}}_{p,2})^{\mathrm{T}} \cdots (\tilde{\mathbf{h}}_{p,M_r})^{\mathrm{T}} \right]^{\mathrm{T}}, \quad (8.34)$$

and using (8.31) for all j, we obtain an equivalent expression

$$\left(\mathbf{I}_{M_r} \otimes \mathcal{D}(\mathbf{X}_p) \right) \mathbf{h}_p = \left(\mathbf{I}_{M_r} \otimes \mathbf{S}_p \right) \tilde{\mathbf{h}}_p. \quad (8.35)$$

Finally, from (8.35) we can simplify (8.29) to

$$\mathbf{y}_p = \sqrt{\rho} \left(\mathbf{I}_{M_r} \otimes \mathbf{S}_p \right) \tilde{\mathbf{h}}_p + \mathbf{w}_p. \quad (8.36)$$

For notation convenience, let us define $\boldsymbol{\mathcal{S}}_p \triangleq (\mathbf{I}_{M_r} \otimes \mathbf{S}_p)$ and $\boldsymbol{\mathcal{V}}_p \triangleq (\mathbf{I}_{M_r} \otimes \mathbf{V}_p)$ such that

$$\boldsymbol{\mathcal{S}}_p = \left(\mathbf{I}_{M_r} \otimes \mathbf{V}_p \right) \boldsymbol{\mathcal{S}}_{p-1} = \boldsymbol{\mathcal{V}}_p \boldsymbol{\mathcal{S}}_{p-1}. \quad (8.37)$$

Accordingly, using (88.36) and (8.37) and after some manipulation, we can write the received signal as

$$\mathbf{y}_p = \boldsymbol{\mathcal{V}}_p \mathbf{y}_{p-1} + \sqrt{2}\,\tilde{\mathbf{w}}_p, \quad (8.38)$$

where $\tilde{\mathbf{w}}_p = (1/\sqrt{2})\mathbf{w}_p - \boldsymbol{\mathcal{V}}_p\mathbf{w}_{p-1}$ is a noise vector each element of which is an independent complex Gaussian random variable with zero mean and unit variance. Without acquiring channel-state information, the detector follows the maximum likelihood decision rule [Hoc01]:

$$\hat{\boldsymbol{\mathcal{V}}}_p = \underset{\boldsymbol{\mathcal{V}}_p \in \mathbb{V}_p}{\mathrm{argmin}} \, \|\mathbf{y}_p - \boldsymbol{\mathcal{V}}_p\mathbf{y}_{p-1}\|_F. \quad (8.39)$$

Even though the decoding complexity increases exponentially with $RKGM_t$, where R is the transmission rate, the decoding complexity can be reduced to a polynomial in KGM_t by a lattice reduction algorithm [Cla01].

8.3 PAIRWISE ERROR PROBABILITY

In this section we provide an approximate PEP formulation based on the results in [Bre01, Sir06b]. We first note that the channel matrix in (8.34) can be reexpressed as

$$\tilde{\mathbf{h}}_p = \tilde{\mathbf{h}}_{p-1} + \Delta\tilde{\mathbf{h}}_p, \quad (8.40)$$

where $\Delta\tilde{\mathbf{h}}_p$ represents the channel mismatch between $\tilde{\mathbf{h}}_p$ and $\tilde{\mathbf{h}}_{p-1}$. For analytical tractability, this section confines the analysis to the case when $\Delta\tilde{\mathbf{h}}_p$ is negligible (i.e., $\tilde{\mathbf{h}}_{p-1} \approx \tilde{\mathbf{h}}_p$). Such a performance formulation provides us with a benchmark for subsequent performance comparisons. Later, in Section 8.4, we show from the numerical results how the channel mismatch affects system performance.

For specific values of T_c and $\tau_{c,l}$, the PEP upper bound is given in [Bre01, proposition 7]. The average PEP can be obtained by averaging over Poisson distributions; however, it is difficult, if not impossible, to obtain the average PEP. In what follows we use the approximation approach as in [Sir06b]. Suppose that $\mathcal{V}_p$ and $\hat{\mathcal{V}}_p$ are two different information matrices. The asymptotic PEP can then be approximated as

$$P_a(\mathcal{V}_p \to \hat{\mathcal{V}}_p) \approx \binom{2\nu-1}{\nu}\left(\prod_{m=1}^{\nu}\beta_{p,m}\right)^{-1}\left(\frac{\rho}{2}\right)^{-\nu}, \tag{8.41}$$

where ρ is an average signal-to-noise ratio per symbol, ν is the rank, and the $\beta_{p,m}$'s are the nonzero eigenvalues of the matrix

$$\mathbf{\Psi}_p \triangleq \mathcal{S}_{p-1}\mathbf{\Sigma}_{\tilde{\mathbf{h}}_p}\mathcal{S}_{p-1}^H(\mathcal{V}_p - \hat{\mathcal{V}}_p)^H(\mathcal{V}_p - \hat{\mathcal{V}}_p), \tag{8.42}$$

in which $\mathbf{\Sigma}_{\tilde{\mathbf{h}}_p} = \mathrm{E}[\tilde{\mathbf{h}}_p\tilde{\mathbf{h}}_p^H]$ denotes the correlation matrix of channel vector $\tilde{\mathbf{h}}_p$.

To simplify the expression for matrix $\mathbf{\Psi}_p$ in (8.42), we evaluate the channel correlation matrix $\mathbf{\Sigma}_{\tilde{\mathbf{h}}_p}$ as follows. Due to the band hopping, the K OFDM symbols in each signal matrix are sent over different subbands. With ideal band hopping, we assume that the signal transmitted over K different frequency bands undergoes independent fading. Assuming also that the MIMO channel is spatially uncorrelated, we can find that $\mathbf{\Sigma}_{\tilde{\mathbf{h}}_p} = \mathbf{I}_{KM_r} \otimes \mathrm{E}[\tilde{\mathbf{h}}_{p,j}^k(\tilde{\mathbf{h}}_{p,j}^k)^H]$, and it can be simplified to

$$\mathbf{\Sigma}_{\tilde{\mathbf{h}}_p} = \mathbf{I}_{KM_r} \otimes \mathrm{diag}\left(\mathbf{R}_{p,1}, \ldots, \mathbf{R}_{p,M_t}\right), \tag{8.43}$$

where $\mathbf{R}_{p,i} \triangleq \mathrm{E}[\tilde{\mathbf{h}}_{p,ij}^k(\tilde{\mathbf{h}}_{p,ij}^k)^H]$ denotes the correlation matrix and is the same for all j's. From (8.33) we can see that the diagonal elements [i.e., the (u, u)th elements, of $\mathbf{R}_{p,i}$] are

$$R_{p,i}^{u,u} = \mathrm{E}[|H_{ij}^k(n_{p,i}^u)|^2] = \mathrm{E}\left[\sum_{c=0}^{C}\sum_{l=0}^{L}\Omega_{c,l}\right] = 1. \tag{8.44}$$

The off-diagonal components [i.e., the (u, v)th for $u \neq v$ components], of $\mathbf{R}_{p,i}$ can be expressed as

$$R_{p,i}^{u,v} = \mathrm{E}\left[H_{ij}^k(n_{p,i}^u)\left(H_{ij}^k(n_{p,i}^v)\right)^H\right]$$
$$= \sum_{c=0}^{C}\sum_{l=0}^{L}\mathrm{E}\left[\Omega_{c,l}e^{-j2\pi\Delta f(n_{p,i}^u - n_{p,i}^v)(T_c+\tau_{c,l})}\right]. \tag{8.45}$$

Observing that $n_{p,i}^u - n_{p,i}^v = u - v$, we can reexpress (8.45) as

$$R_{p,i}^{u,v} = \Omega_{0,0} \sum_{c=0}^{C} \sum_{l=0}^{L} \mathrm{E}\left[e^{-g(1/\Gamma,u,v)T_c - g(1/\gamma,u,v)\tau_{c,l}} \right], \tag{8.46}$$

where $g(a, u, v) = a + \mathbf{j}2\pi(u - v)\Delta f$. According to the Poisson distribution of the multipath delays, T_c and $\tau_{c,l}$ can be modeled as summations of iid exponential random variables with parameters Λ and λ, respectively. Therefore, averaging (8.46) over the distribution of T_c and $\tau_{c,l}$, we arrive at

$$R_{p,i}^{u,v} = \Omega_{0,0} \sum_{c=0}^{C} \sum_{l=0}^{L} \frac{\Lambda + g(1/\Gamma, u, v)}{g(1/\Gamma, u, v)} \frac{\lambda + g(1/\gamma, u, v)}{g(1/\gamma, u, v)}. \tag{8.47}$$

Since $R_{p,i}^{u,v}$ is the same for all i's and p's, we denote $\mathbf{R} \triangleq \mathbf{R}_{p,i}$, which allows us to simplify (8.43) further to

$$\Sigma_{\tilde{\mathbf{h}}_p} = \mathbf{I}_{KM_tM_r} \otimes \mathbf{R}. \tag{8.48}$$

Substituting (8.48) into (8.42) and applying the property of the tensor product $(\mathbf{A}_1 \otimes \mathbf{B}_1)(\mathbf{A}_2 \otimes \mathbf{B}_2)(\mathbf{A}_3 \otimes \mathbf{B}_3) = (\mathbf{A}_1\mathbf{A}_2\mathbf{A}_3 \otimes \mathbf{B}_1\mathbf{B}_2\mathbf{B}_3)$, we obtain

$$\Psi_p = \mathbf{I}_{M_r} \otimes \Theta_p, \tag{8.49}$$

in which

$$\Theta_p = \mathbf{S}_{p-1}\left(\mathbf{I}_{KM_t} \otimes \mathbf{R}\right)\mathbf{S}_{p-1}^H \Delta, \tag{8.50}$$

and $\Delta = (\mathbf{V}_p - \hat{\mathbf{V}}_p)^H(\mathbf{V}_p - \hat{\mathbf{V}}_p)$. Hence, by (8.49), the PEP in (8.41) can be expressed as

$$P_a(\mathbf{V}_p \to \hat{\mathbf{V}}_p) \approx \binom{2rM_r - 1}{rM_r} \left(\prod_{m=1}^{r} \lambda_{p,m}\right)^{-M_r} \left(\frac{\rho}{2}\right)^{-rM_r}, \tag{8.51}$$

where r is the rank of Θ_p and $\lambda_{p,m}$'s are the nonzero eigenvalues of Θ_p.

To quantify the maximum diversity order, which is the exponent of $\rho/2$ in (8.51), we first note that $\mathbf{S}_{p-1}$ and $\mathbf{V}_p$ are of size $KGM_t \times KGM_t$, and the correlation matrix $\mathbf{R}$ is of size $G \times G$. Therefore, the maximum diversity gain is

$$G_d^{\max} = M_r \max\left(\min_{\forall \mathbf{V}_p^k \neq \hat{\mathbf{V}}_p^k} \mathrm{rank}(\Theta_p) \right) = KGM_tM_r. \tag{8.52}$$

Note that $\mathbf{R}$ is of full rank if G is less than the total number of multipath components $(C + 1)(L + 1)$. Due to the large bandwidth of a UWB waveform, the signal received typically contains a significant number of resolvable multipath components. Consequently, the correlation matrix $\mathbf{R}$ is generally of full rank. Therefore, the maximum diversity order of KGM_tM_r can be achieved by using a set of proper designed codeword matrices $\mathbf{V}_p$.

When the maximum diversity order is achieved, we have

$$\prod_{m=1}^{KGM_t} \lambda_{p,m} = \det\left(\mathbf{S}_{p-1}(\mathbf{I}_{KM_t} \otimes \mathbf{R})\mathbf{S}_{p-1}^H\right) \times \det\left((\mathbf{V}_p - \hat{\mathbf{V}}_p)^H(\mathbf{V}_p - \hat{\mathbf{V}}_p)\right). \quad (8.53)$$

By applying the identity $\det(\mathbf{AB}) = \det(\mathbf{BA})$ and using the unitary property of matrix $\mathbf{S}_p^{k-1}$, the coding gain can be determined as [Hoc01]

$$\zeta = \frac{1}{2} \min_{\mathbf{V}_p^k \neq \hat{\mathbf{V}}_p^k} \left| \prod_{m=1}^{KGM_t} \lambda_{p,m} \right|^{1/2KGM_t},$$

$$= \frac{1}{2} \left| \prod_{i=1}^{M_t} \det(\mathbf{R}) \right|^{1/2KGM_t} \min_{\mathbf{V}_p \neq \hat{\mathbf{V}}_p} \prod_{m=1}^{M} \left| v_{p,m} - \hat{v}_{p,m} \right|^{1/KGM_t}. \quad (8.54)$$

The results in (8.52) and (8.54) lead to some interesting observations as follows. From (8.54), the differential UWB multiband OFDM system achieves the same diversity gain under a different channel environment. This implies that the clustering property of a UWB channel does not strongly affect the diversity gain of a differential multiband system. Moreover, by incorporating the frequency-domain differential scheme with the multiband OFDM transmission, the scheme is able to achieve the diversity gain of GKM_tM_r, regardless of the channel time-correlation property. This is different from the use of differential STF coding in conventional MIMO-OFDM systems (e.g., in [Wan02]), where the maximum achievable diversity gain is only GM_tM_r, due to the requirement of almost constant channels over several OFDM blocks. Furthermore, (8.54) reveals that the coding gain is severely affected by the multipath arrival rates and decay factors through the correlation matrix $\mathbf{R}$.

8.4 SIMULATION RESULTS

We performed simulations for a multiband OFDM system with $N = 128$ subcarriers and each subband occupies a bandwidth of 528 MHz. The channel model parameters followed those for CM1 and CM2 [Foe03b]. The data matrix $\mathbf{V}_p$ in (8.16) were constructed by joint coding across G, K, and M_t using existing cyclic group codes [Hoc01]. In case of repetition-based coding, the codeword is given by $\mathbf{V}_p = \mathbf{I}_K \otimes \mathbf{v}_p$, where $\mathbf{v}_p$ is a $GM_t \times GM_t$ jointly encoded diagonal matrix.

Figure 8.3 depicts the performances of a single-antenna multiband OFDM system with different numbers of G and K. For a fair comparison, the spectral efficiency is fixed at $R = 1$ bit/s · Hz for all cases. The performances are simulated under CM1. For an uncoded differential system ($G = 1$ and $K = 1$), we can see that the performance loss is more than 3 dB compared to the coherent detection, and an error floor can be observed. This is due to the effect of the channel mismatch between adjacent subcarriers. By jointly encoding across two OFDM symbols ($G = 1$ and $K = 2$), the diversity gain is increased, hence resulting in significant performance improvement.

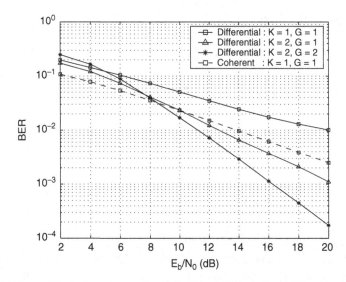

Figure 8.3 Performance under CM1, $M_t = 1$, $M_r = 1$, $R = 1$ bit/s $\cdot$ Hz.

As shown in Fig. 8.3, the performance gain is more than 7 dB at a BER of 10^{-2}. By further joint encoding across two subcarriers ($G = 2$ and $K = 2$), the DMB-OFDM scheme obtains an additional 4 dB of gain at a BER of 10^{-3}. This observation is in accordance with our theoretical result in (8.51) that the performance can be improved by increasing the number of jointly encoded subcarriers or the number of jointly encoding OFDM symbols. Moreover, at a high SNR, the jointly encoding differential scheme outperforms the uncoded multiband OFDM system with coherent detection. We observe about 1 to 2 dB of gain when $G = 1$ and $K = 2$ and about 3 to 5 dB of gain when $G = 2$ and $K = 2$ at BER between 10^{-2} and 10^{-3}.

In Fig. 8.4 we compare the performance of the DMB-OFDM scheme under CM1 and CM2. The information is transmitted repeatedly across $K = 1$, 2, and 3 OFDM symbols; hence the transmission rate is $1/K$ bit/s $\cdot$ Hz. We can see that the performance of the DMB-OFDM scheme under CM1 is better than that under CM2 for all cases. This is due to the fact that the multipath components in CM2 are more random than those in CM1, which implies that compared with CM1, CM2 results in larger channel mismatch and hence worse performance. For each channel model, the performance improves as the number of encoded OFDM symbols increases, which confirms our theoretical analysis.

Figure 8.5 depicts the performances of differential multiband OFDM systems. The number of jointly encoded OFDM symbols is fixed at $K = 1$, and the spectral efficiency is $R = 1$ bit/s $\cdot$ Hz for all cases. From Fig. 8.5 we can observe the performance improvement as the number of antennas increases. When using two transmitting and one receiving antennas and encoding across one subcarrier and one OFDM symbol, the DMB-OFDM scheme yields a 7-dB improvement over the single-antenna system.

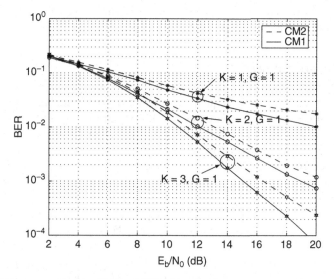

Figure 8.4 Performance under CM1 and CM2: $M_t = 1$, $M_r = 1$, $R = 1$ and $R = 1/K$ bit/s · Hz.

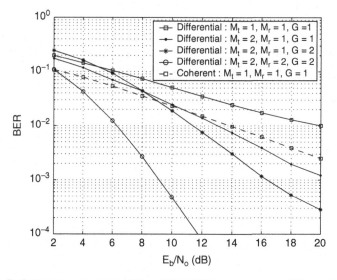

Figure 8.5 Performance comparison of the DMB-OFDM scheme under CM1 employing SISO and MIMO processing; $K = 1$ and $R = 1$ bit/s · Hz.

When we jointly encode across two subcarriers, an additional performance gain of about 4 dB can be observed at a BER of 10^{-3}. However, slight error floors can still be observed when the data are encoded across multiple transmitting antennas since the chance of channel mismatch is higher in this case. On the other hand, increasing the number of receiving antennas improves the diversity gain without a trade-off in channel mismatch. In particular, an additional performance gain of 6 dB is observed when two receiving antennas are employed.

8.5 CHAPTER SUMMARY

We present in this chapter a frequency-domain differential modulation scheme for multiband OFDM systems, called a DMB-OFDM scheme. By a technique of band hopping in combination with joint coding across spatial, temporal, and frequency domains, the scheme proposed is able to explore available spatial and multipath diversities, richly inherent in UWB environments. The analysis reveals that the differential scheme proposed achieves the same diversity advantage under different channel environments. However, the scheme proposed obtains different coding gains, which depends on the clustering behavior of UWB channels. In particular, higher coding gain is achieved in a shorter-range scenario. For a single-antenna UWB multiband OFDM system, simulation results show that the differential multiband scheme proposed results in a performance superior to the conventional differential encoding scheme, particularly under a short-range line-of-sight scenario (e.g., in CM1). We obtain about 7 dB of gain at a BER of 10^{-2} when encoding jointly across one subcarrier and two OFDM symbols. Moreover, at a high SNR range, the jointly encoded differential scheme proposed outperforms the uncoded coherent detection scheme by about 3 to 5 dB at a BER between 10^{-2} and 10^{-3}. In case of a UWB multiband OFDM system with multiple transmitting antennas, while a slight error floor occurs due to the effect of channel mismatch, additional diversity can be observed when the number of transmitting antennas is increased. However, increasing the number of receiving antennas improves diversity gain while reducing the effect of channel mismatch.

9

POWER-CONTROLLED
CHANNEL ALLOCATION

For a UWB device to coexist with other devices, the transmitter power level of UWB is strictly limited by the FCC spectral mask. Such a limitation poses a significant design challenge to any UWB system. Efficient management of the limited power is thus a key feature to fully exploit the advantages of UWB. The low transmitting power of UWB emissions ensures a long lifetime for the energy-limited devices. In addition, UWB systems are expected to support an integration of multimedia traffic, such as voice, image, data, and video streams. This requires a cross-layer algorithm that is able to allocate the available resources to a variety of users with different service rates in an effective way.

Most of the existing resource allocation schemes for UWB systems (see [Zhu03, Cuo02, Rad04] and references therein) are based on single-band impulse-radio technology. On the other hand, most research efforts on multiband UWB systems have been devoted to issues related to the physical layer [Bat04, Sab04, Nak04]. Some of the key issues in multiband UWB systems that remain largely unexplored are resource allocations such as power control and channel allocation. As presented in Chapter 4, the current multiband OFDM proposal divides the subbands into groups, each comprising two or three subbands. A set of certain time–frequency codes is used to interleave the data within each band group [Bat04]. This strategy lacks the ability to allocate subbands optimally since the subbands available are not assigned to each user according to its channel condition. Moreover, in the multiband OFDM proposal [Bat04], the transmitting power of each user is distributed equally among its assigned subbands without any power adaptation to the channel variations. So adaptive optimization of the subband assignment and power control can greatly improve the system performances of UWB multiband OFDM systems.

In this chapter we present a power-controlled channel allocation (PCCA) scheme for UWB multiband OFDM wireless networks [Sir05b, Sir07]. By allocating the

Adapted with permission from © 2007 *IEEE Transactions on Wireless Communications*, Vol. 6, no. 2, Feb. 2007, pp. 583–592.

subbands, transmitting power, and data rates efficiently among all users, the scheme enables the multiband OFDM system to operate at a low transmitter power level while still achieving the performance desired. First, we formulate a subband assignment and power allocation problem as an optimization problem whose goal is to minimize the overall transmitting power provided that all users achieve their requested data rates and desired packet error rate (PER), while the power spectral density complies with the FCC limit [FCC02]. To take into account the fact that users in the multiband OFDM system may have different data rates, which in turn implies different channel coding rates frequency spreading gains, and/or time spreading gains, our formulated problem considers not only the limitation on transmitter power level, but also band hopping for users with different data rates. It turns out that the problem formulated is an integer programming problem whose complexity is NP hard. Then, to reduce the complexity of the problem formulated, we present a fast suboptimal algorithm that can guarantee that we will obtain a nearly optimal solution but requires low computational complexity. To ensure system feasibility in variable channel conditions, we further develop a joint rate assignment and power-controlled channel allocation algorithm that is able to allocate resources to users according to three different system optimization goals: maximizing overall rate, achieving proportional fairness, and reducing maximal rate. Simulation results based on the UWB channel model specified in the IEEE 802.15.3a standard [Foe03b] show that the PCCA scheme achieves up to 61% of transmitter power saving compared to a standard multiband OFDM scheme [Bat04]. Moreover, the PCCA scheme can also find feasible solutions adaptively when the initial system is not feasible for the rate requirements of the users.

In Section 9.1 we describe the system model of multiband OFDM. In Section 9.2 we first formulate the power-controlled channel allocation problem. Then a fast suboptimal scheme is developed. Finally, we present a joint rate assignment and resource allocation algorithm to ensure system feasibility. Simulation results are given in Section 9.3.

9.1 SYSTEM MODEL

We consider a UWB system using multiband OFDM that has been presented for the IEEE 802.15.3a WPAN standard [TG3a]. Let the available UWB spectrum be divided into S subbands, each occupying a bandwidth of at least 500 MHz, in compliance with FCC regulations. The UWB system employs OFDM with N subcarriers, which are modulated using QPSK. At each OFDM symbol period, the modulated symbol is transmitted over one of the S subbands. These symbols are time-interleaved across subbands. Different data rates are achieved using different channel coding, frequency spreading, or time spreading rates [Bat04].

As described in Chapter 4, frequency-domain spreading is obtained by choosing conjugate symmetric inputs to the IFFT. Specifically, $N/2$ complex symbols are transmitted in the first half of the subcarriers, and their conjugate symmetric symbols are transmitted in the second half of the subcarriers. The time-domain spreading is achieved by transmitting one OFDM symbol followed by a permutation of that

OFDM symbol (i.e., transmitting the same information twice during two OFDM symbol periods). The receiver combines the information transmitted via different times or frequencies to increase the SNR of data received.

The channel model is based on the S-V model. It is worth noting that for most WPAN applications, the transmitter and receiver are stationary [Mol03]. As a result, a UWB channel is very slowly fading. The standard channel model assumes that the channel stays either completely static or is time invariant during the transmission of each packet [Foe03b, Mol03]. We assume that the channel-state information is known at both the transmitter and the receiver.

We consider a multiuser scenario where K users transmit their information simultaneously. The kth user has the data rate R_k, which can be any value specified in Table 4.2. Multiple access is enabled by the use of suitably designed band-hopping sequences. Such hopping sequences are based on time-frequency codes [Bat04], which are assigned uniquely to different users so as to enable frequency diversity while minimizing multiuser interference. As shown in Table 4.2, if the rate is higher than 200 Mbps, there is no time spreading; otherwise, the time-domain spreading operation is performed with a spreading factor of 2. In this case, any time–frequency code with a period of 2 can guarantee that each user will achieve the additional diversity by transmitting the same information over two OFDM blocks. Time–frequency codes with period longer than 2 can also be used to improve the multiple access capability for asynchronous UWB wireless networks [Bat04]. To simplify the problem formulation, we consider here a multiband OFDM system employing time–frequency codes of length 2. The extension to UWB systems with longer time–frequency codes is straightforward.

To specify in which subbands each user can transmit its information, we define a $K \times S$ assignment matrix $\mathbf{A}$, whose (k, s)th element is denoted by a_{ks}, for $k = 1, 2, \ldots, K$ and $s = 1, 2, \ldots, S$. This a_{ks} represents the number of OFDM symbols that user k is allowed to transmit on the sth subband during two OFDM symbol periods. Assuming that each user utilizes one subband per transmission, a_{ks} can take any value from the set $\{0, 1, 2\}$. However, when the kth data rate of the user is less than or equal to 200 Mbps, we need to ensure that band hopping is performed to obtain the diversity from time spreading. In this case, a_{ks} is restricted to $a_{ks} \in \{0, 1\}$. Thus, the elements of the assignment matrix satisfy [Sir05b]

$$a_{ks} \in \phi(R_k) = \begin{cases} \{0, 1\}, & R_k \leq 200\,\text{Mbps}; \\ \{0, 1, 2\}, & R_k > 200\,\text{Mbps}. \end{cases} \tag{9.1}$$

During each OFDM symbol period, one user will occupy one subband. Since we consider the duration of two OFDM blocks, the assignment strategy needs to satisfy

$$\sum_{s=1}^{S} a_{ks} = 2, \quad k = 1, 2, \ldots, K. \tag{9.2}$$

In addition, to minimize the multiple access interference, each subband is assigned to a specific user at a time, and hence each subband can be used at most twice during two OFDM symbol periods. Therefore, the subband assignment follows

$$\sum_{k=1}^{K} a_{ks} \leq 2, \quad s = 1, 2, \ldots, S. \tag{9.3}$$

Let $P_k^s(n)$ denote the transmitting power of the kth user at subcarrier n of the sth subband. Accordingly, the SNR of user k at the sth subband and the nth subcarrier is given by

$$\Gamma_k^s(n) = \frac{P_k^s(n)G_k^s(n)}{\sigma_k^2}, \tag{9.4}$$

where $G_k^s(n)$ is the corresponding channel gain. We can express $G_k^s(n)$ as

$$G_k^s(n) = \left|H_k^s(n)\right|^2 \left(\frac{4\pi d_k}{\lambda_k^s}\right)^{-\nu}, \tag{9.5}$$

in which $H_k^s(n)$ is the channel frequency response at subband s and subcarrier n, ν is the propagation loss factor, d_k represents the distance between the transmitter and the receiver, $\lambda_k^s = 3 \times 10^8 / f_{c,k}^s$ is the wavelength of the signal transmitted, and $f_{c,k}^s$ is the center frequency of the waveform. In (9.4), σ_k^2 denotes the noise power at each subcarrier, which is defined as

$$\sigma_k^2 = 2 \times 10^{(-174+10\log_{10}(R_k)+N_F)/10}, \tag{9.6}$$

where R_k is the data rate of the kth user and N_F is the received noise figure referred to the antenna terminal [Bat04]. As in the multiband standard proposal, we assume that the noise power σ_k^s is the same for every subcarrier within each subband. We assume ideal band hopping such that signals transmitted over different subbands undergo independent fading and there is no multiple access interference.

Due to consideration for the simple transceiver of UWB, the current standard assumes that there is no bit loading and that the power is equally distributed across subcarriers within each subband. Similarly, we assume that $P_k^s(n) = P_k^s(n')$ for any $0 \leq n, n' \leq N - 1$. Denote

$$P_k^s(n) = P_k^s, \quad n = 0, 1, \ldots, N - 1; \tag{9.7}$$

then the $K \times S$ power allocation matrix can be defined as $[\mathbf{P}]ks = P_k^s(n)$, in which the (k, s)th component represents the transmitting power of the kth user in subband s.

9.2 POWER-CONTROLLED CHANNEL ALLOCATION SCHEME

In the multiband OFDM frequency band plan [Bat04], the subband center frequencies span a wide range, from 3.43 to 10.3 GHz. Consequently, different subbands tend to undergo different fading and propagation loss. Additionally, the channel condition for a specific subband may be good for more than one user. Therefore, to reduce the power consumption efficiently, we need to optimize the subband assignment matrix $\mathbf{A}$ and power allocation matrix $\mathbf{P}$ under some practical constraints.

In this section, first, we derive a generalized SNR expression for various UWB transmission modes. Second, we provide a necessary condition for the SNR so as to satisfy the PER requirement. Then we present a problem formulation to minimize the overall transmitter power provided that all users achieve their requested data rates and desired PER, while the transmitter power level is below the FCC limitation, and rate parameters follow the standard proposal given in Table 4.2. We develop a fast suboptimal PCCA scheme to solve the optimization problem. Finally, to ensure system feasibility, we develop a joint rate adaptation, subband assignment, and power allocation algorithm.

9.2.1 Generalized SNR for Various Transmission Modes

Assuming that the channel state information is known perfectly at the receiver, the receiver employs a maximum ratio combiner (MRC) to combine the information transmitted via different times or frequencies. As a result, the average SNR at the output of MRC depends not only on the channel coding rate but also on the time and frequency spreading factors. The following proposition provides a generalized expression of the average SNR for any data rates.

Proposition. Assume maximum ratio combining and $P_k^s(n) = P_k^s$ for all subcarriers n; then the average SNR of the kth user is given by

$$\bar{\Gamma}_k = \sum_{s=1}^{S} a_{ks} P_k^s F_k^s, \tag{9.8}$$

where

$$F_k^s \triangleq \frac{b_k}{N\sigma_k^2} \sum_{n=0}^{N-1} G_k^s(n) \tag{9.9}$$

and b_k is a constant that depends on the data rate of the kth user as follows:

$$b_k = \begin{cases} 2, & R_k \leq 80\,\text{Mbps}; \\ 1, & 80 < R_k \leq 200\,\text{Mbps}; \\ 1/2, & R_k > 200\,\text{Mbps}. \end{cases} \tag{9.10}$$

PROOF: Recall that when R_k is less than 80 Mbps, the information is spread across both time and frequency with an overall spreading gain of 4. Consequently, the total SNR for the kth user at subcarrier n, $n = 0, 1, \ldots, N/2 - 1$, is

$$\Gamma_k(n) = \sum_{s=1}^{S} a_{ks} \left[\Gamma_k^s(n) + \Gamma_k^s(n + N/2) \right]. \tag{9.11}$$

Note that the SNR in (9.11) is based on the assumption that there is no multiuser interference and no correlation among the data bits; it leads to an upper bound on the performance. Averaging (9.11) over $N/2$ subcarriers results in the average SNR:

$$\bar{\Gamma}_k = \frac{1}{N/2} \sum_{n=0}^{N/2-1} \Gamma_k(n) = \frac{1}{N/2} \sum_{n=0}^{N-1} \sum_{s=1}^{S} a_{ks} \Gamma_k^s(n). \tag{9.12}$$

By substituting (9.4) into (9.12) and assuming that $P_k^s(n) = P_k^s$, we obtain

$$\bar{\Gamma}_k = \frac{2}{N} \sum_{n=0}^{N-1} \sum_{s=1}^{S} a_{ks} P_k^s \frac{G_k^s(n)}{\sigma_k^2} = \sum_{s=1}^{S} a_{ks} P_k^s \left(\frac{2}{N\sigma_k^2} \sum_{n=0}^{N-1} G_k^s(n) \right). \tag{9.13}$$

When R_k is between 106.7 and 200 Mbps, only time spreading is performed, and hence the total SNR at subcarrier n, $n = 0, 1, \ldots, N - 1$, becomes

$$\Gamma_k(n) = \sum_{s=1}^{S} a_{ks} \Gamma_k^s(n) = \sum_{s=1}^{S} a_{ks} \frac{P_k^s(n) G_k^s(n)}{\sigma_k^2}. \tag{9.14}$$

Thus, the average SNR can be obtained from (9.14) as

$$\bar{\Gamma}_k = \frac{1}{N} \sum_{n=0}^{N-1} \Gamma_k(n) = \sum_{s=1}^{S} a_{ks} P_k^s \left(\frac{1}{N\sigma_k^2} \sum_{n=0}^{N-1} G_k^s(n) \right). \tag{9.15}$$

For R_k higher than 200 Mbps, there is no spreading and the average SNR of the kth user is simply the average of $\Gamma_k^s(n)$ over N subcarriers and two subbands:

$$\bar{\Gamma}_k = \frac{1}{2N} \sum_{n=0}^{N-1} \sum_{s=1}^{S} a_{ks} \Gamma_k^s(n) = \sum_{s=1}^{S} a_{ks} P_k^s \left(\frac{1}{2N\sigma_k^2} \sum_{n=0}^{N-1} G_k^s(n) \right). \tag{9.16}$$

Express (9.13), (9.15), and (9.16) in terms of F_k^s defined in (9.9), leading to the results in (9.8).

9.2.2 PER and Rate Constraint

A common performance requirement of UWB systems is to offer packet transmission with an error probability below a desired threshold value. The PER metric is related directly to BER performance, which in turn depends on the SNR at the output of the MRC. By keeping the SNR level higher than a specific value, the PER can be ensured to be lower than the PER threshold. In the sequel we provide a necessary condition for the average SNR so as to satisfy the PER requirement.

Suppose that the maximum PER is ε and the packet length is L bits; then the bit error probability after the channel decoder for the kth user, $\mathcal{P}_k$, needs to satisfy

$$1 - (1 - \mathcal{P}_k)^L \le \varepsilon. \tag{9.17}$$

By the assumptions of the use of convolutional coding and Viterbi decoding with perfect interleaving, $\mathcal{P}_k$ is given by [Pro01]

$$\mathcal{P}_k \leq \sum_{d=d_{\text{free}}}^{\infty} a_d \mathcal{P}_k(d), \tag{9.18}$$

where d_{free} is the free distance of the convolutional code, a_d denotes the total number of error events of weight d, and $\mathcal{P}_k(d)$ represents the probability of choosing the incorrect path with distance d from the correct path. Assume hard-decision decoding; then $\mathcal{P}_k(d)$ is related to the average BER, $\bar{B}_k$, as [Pro01]

$$\mathcal{P}_k(d) = \sum_{l=(d+1)/2}^{d} \binom{d}{l} \bar{B}_k^l (1 - \bar{B}_k)^{d-l} \tag{9.19}$$

when d is odd, and

$$\mathcal{P}_k(d) = \sum_{l=d/2+1}^{d} \binom{d}{l} \bar{B}_k^l (1 - \bar{B}_k)^{d-l} + \frac{1}{2}\binom{d}{d/2} \bar{B}_k^{d/2}(1 - \bar{B}_k)^{d/2} \tag{9.20}$$

when d is even.

The average BER, $\bar{B}_k$, can be obtained by averaging the conditional BER over the probability density function of the SNR at the output of MRC. With Γ_k denoting the instantaneous SNR at the MRC output, the conditional BER is given by [Pro01]

$$B_k(\Gamma_k) = Q(\sqrt{\Gamma_k}), \tag{9.21}$$

where $Q(\cdot)$ is the Gaussian error function. From (9.17) and (9.18) we can see that for a given value of PER threshold ε, a corresponding BER threshold can be obtained. Since the error probability $\mathcal{P}_k$ in (9.18) is related to the coding rate through the parameters d_{free} and a_d, the BER requirement depends not only on the value of ε, but also on the data rate R_k. This implies that the SNR threshold is also a function of both ε and R_k. Let $\gamma(\varepsilon, R_k)$ be the minimum SNR of the kth user that is required to achieve the data rate R_k with PER less than ε. Then the necessary condition for the average SNR [defined in (9.8)] to satisfy the PER requirement is given by

$$\bar{\Gamma}_k = \sum_{s=1}^{S} a_{ks} P_k^s F_k^s \geq \gamma(\varepsilon, R_k). \tag{9.22}$$

9.2.3 Problem Formulation

The optimization goal is to minimize the overall transmitter power subject to the PER, rate, and FCC regulation constraints. Recall from (9.1) that the assignment matrix **A** has $a_{ks} \in \phi(R_k)$, $\forall k, s$. We can formulate the problem as follows:

$$\min_{\mathbf{A},\mathbf{P}} P_{\text{sum}} = \sum_{k=1}^{K} \sum_{s=1}^{S} a_{ks} P_k^s \tag{9.23}$$

$$\text{s.t.} \begin{cases} \text{Rate and PER: } \sum_{s=1}^{S} a_{ks} P_k^s F_k^s \geq \gamma(\varepsilon, R_k), \ \forall k \\ \text{Assignment (9.2): } \sum_{s=1}^{S} a_{ks} = 2, \ \forall k \\ \text{Assignment (9.3): } \sum_{k=1}^{K} a_{ks} \leq 2, \ \forall s \\ \text{Power: } P_k^s \leq P_{\max}, \ \forall k, s, \end{cases}$$

where the first constraint in (9.23) is to ensure rate and PER requirements. The second and third constraints are described in Section 9.1. The last constraint is related to the fact that the transmitter power spectral density is limited to -41.3 dBm/MHz, according to FCC Part 15 rules [FCC02]. Here $P_{\max}$ is the maximum transmitting power, taking into consideration effects such as peak-to-average power ratio (PAPR) (i.e., $P_{\max}$ is the maximum power considering both the average maximum power allowed by the FCC and the PAPR of OFDM signals).

If the elements in the assignment matrix $\mathbf{A}$ are binary, the problem defined in (9.23) can be viewed as a generalized form of the generalized assignment problem [Kel04], which is NP hard. Since the components of $\mathbf{A}$ can be 0, 1, or 2, the problem is an even more difficult integer programming problem. So the existing channel assignment approaches (e.g., in [Won99]) are not applicable in (9.23). Although the optimal solution can be found through a full search, it is computationally expensive. To overcome the complexity issue, in Section 9.2.4 we present a fast suboptimal scheme, which is nearly optimal but has very low computational complexity.

9.2.4 Subband Assignment and Power Allocation Algorithm

The basic idea is a greedy approach to assign a_{ks} for a user step by step so that the power consumption is minimized. The initialization is to set $\mathbf{A} = \mathbf{0}_{K \times S}$, define the user optimization list $K_{\text{live}} = \{1, 2, \ldots, K\}$, and define the subband optimization list $S_{\text{live}} = \{1, 2, \ldots, S\}$. First, each user hypothesizes that it can assign its transmission to different subbands regarding the absence of other users. For each hypothesis, a dummy overall transmission power P_{dummy}^k is calculated by finding the minimum power among all possible subbands such that the BER performance requirement of user k is satisfied. The user with the highest dummy overall transmitting power to achieve its rate will be assigned first, so that the best channel is assigned to the user that can reduce the overall power most. Then this user is removed from the optimization list, K_{live}. Since each subband can accommodate only one user per symbol period and we consider two OFDM symbol periods, when a subband is assigned twice, this subband is removed from the optimization list S_{live}. Then we go to the first step for the rest of the users, to assign their transmissions to the rest of the subbands. This iteration is continued until all users are assigned with their subbands (i.e., $K_{\text{live}} = \emptyset$. Finally, the algorithm checks if the maximum power is larger than the power limitation. If yes, an outage is reported; otherwise, the final values of $\mathbf{A}$ and $\mathbf{P}$ are obtained. The algorithm can be described as follows:

Initialization: $a_{ks} = 0, \ \forall k, s, K_{\text{live}} = \{1, \ldots, K\}, S_{\text{live}} = \{1, \ldots, S\}$

Iteration: Repeat until $k_{\text{live}} = \emptyset$ or $S_{\text{live}} = \emptyset$

1. For $k \in K_{\text{live}}$

$$P_{\text{dummy}}^k = \min \sum_{s=1}^{S} a_{ks} P_k^s \text{ s.t. } a_{ks} \in S_{\text{live}}$$

   End
2. Select k' with the maximal P_{dummy}^k, $\forall k$, assign the corresponding a_k's to **A**, and update **P**.
3. $K_{\text{live}} = K_{\text{live}} \backslash k'$.
4. If $\sum_{k=1}^{K} a_{ks'} = 2$, $S_{\text{live}} = S_{\text{live}} \backslash s'$, $\forall s'$.

   End: If $(\max(\mathbf{P}) > P_{\max})$ or $(S_{\text{live}} = \emptyset$ and $K_{\text{live}} \neq \emptyset)$, an outage is reported. Otherwise, return **A** and **P**.

The complexity of the algorithm is only $O(K^2 S)$. Although the algorithm is suboptimal, simulation results illustrated in the succeeding section show that the performance of the fast suboptimal algorithm is very close to that of the optimal solutions obtained by full search. Another complexity issue is that in this PCCA scheme, power control is needed for each subband.[1] This will increase the system complexity slightly, but from the simulation results, we can see that the performance improvement is significant. Moreover, the algorithm can be implemented by the master node to manage the power and subband use of all users in a UWB piconet system, as adopted in the IEEE 802.15.3a standard [TG3a]. The signaling information that needs to be broadcast at the master node includes the band-hopping sequence of each user and the corresponding transmitting power. The algorithm is updated when a new user joins the network or when the channel link quality of each user changes considerably. In each update, the algorithm requires the channel-state information for all subbands considered (instead of three subbands as in the standard multiband scheme) between every transmitter and the receiver. Such an update does not occur frequently, thanks to the small size of the piconet and the stationary nature of most transceivers in WPAN applications.

9.2.5 Joint Rate Assignment and Resource Allocation Algorithm

Since the transmitting power in each subband is limited by maximal power $P_{\max}$, solutions to (9.23) may not exist in some situations, such as when the user rates are high but the channel conditions are poor. Under such conditions, some user rates are set too high, and we call the system unworkable. In this case the rates requested need to be lowered. Here we develop a joint rate assignment and power-controlled channel allocation algorithm that is able to arrive at feasible solutions adaptively when the user rates set for the initial system are too high. Basically, the algorithm comprises two

---

[1] But no power control or bit loading is required for subcarriers within each subband.

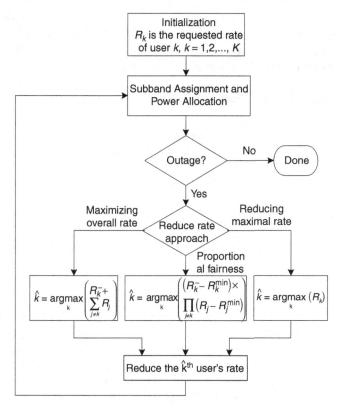

Figure 9.1 Flowchart of the joint rate assignment and resource allocation algorithm.

main stages: resource allocation and rate adaptation stages. Figure 9.1 is a flowchart of the algorithm.

At the initialization step, the data rate of the kth user, R_k, $k = 1, 2, \ldots, K$, is set to the rate requested. After the initial setting, the first stage is to perform the subband and power allocation using the algorithm described in Section 9.2.4. If there is a solution for this assignment, it is done. Otherwise, an outage will be reported, indicating that the rates requested of users are too high for current channel conditions. In this case we proceed to the second stage, where the rate adaptation is performed.

In the rate adaptation stage, the algorithm chooses only one user, reducing its rate to the next lower rate as listed in Table 4.2. To specify which user is to be selected, we consider three goals: maximizing the overall rate, achieving proportional fairness [Kel97], [2] and reducing the maximal rate. In particular, given the data rate of the kth user, R_k, we denote its one-step reduced rate by R_k^-. For instance, from Table 4.2 the

---

[2]Note that proportional fairness is achievable when the utility is a log function. Here we have a discrete and nonconvex case, so the same product form is used as the system performance goal instead of the log function. From the simulations, this goal achieves trade-off of performances and fairness between the maximal rate goal and reducing maximal rate goal..

reduced rate R_k^- corresponding to a rate $R_k = 320$ Mbps is $R_k^- = 200$ Mbps. Note that when the rate R_k reaches the minimum allowable rate of 53.3 Mbps, we let $R_k^- = R_k$ (i.e., the rate R_k is not reduced further). Then the user $\hat{k}$ whose rate will be reduced can be determined according to the performance goals as:

- Maximizing the overall rate:

$$\hat{k} = \underset{k}{\operatorname{argmax}} \, R_k^- + \sum_{j=1, j \neq k}^{K} R_j.$$

- Achieving proportional fairness:

$$\hat{k} = \underset{k}{\operatorname{argmax}} \, \prod_{j=1, j \neq k}^{K} \left(R_j - R_j^{\min} \right) \times \left(R_k^- - R_k^{\min} \right).$$

- Reducing the maximal rate:

$$\hat{k} = \underset{k}{\operatorname{argmax}}(R_k).$$

Here $R_k^{\min}$ denotes a minimal rate requirement for user k. With the maximizing overall rate approach, the overall system rate is maximized in every reduction step. In the proportional fairness approach, the product of rates minus minimal rate requirements [Kel97] is maximized. For the reducing maximal rate approach, the highest rate in the system will be reduced. Note that if there is still no solution to the assignment after the rates of all users are reduced to the minimum allowable rate, an outage is reported. This indicates that under current channel conditions, the system cannot support the transmission of all K users at the same time. The joint resource allocation and rate adaptation algorithm is summarized as follows.

Initialization: Iteration index $n' = 0$, $R_k(0) = $ requested rate of user k, $k = 1, 2, \ldots, K$

Iteration:

1. Given $R_k(n')$, solve the subband assignment and power allocation problem in (9.23).
2. If (9.23) has a solution, the algorithm ends. Otherwise:
   - If $R_k(n') = R_k^-(n')$, $\forall k$, an outage is reported and the algorithm ends.
   - Determine $\hat{k}$.
   - Update the rates:

   $$R_k(n' + 1) = \begin{cases} R_k^-(n'), & k = \hat{k}; \\ R_k(n'), & \text{otherwise.} \end{cases}$$

   - Set $n' = n' + 1$.

9.3 SIMULATION RESULTS

We perform simulations for UWB multiband OFDM systems with $N = 128$ subcarriers, $S = 14$ subbands, and the subband bandwidth of 528 MHz. Following the IEEE 802.15.3a standard proposal [Bat04], we utilize subbands with center frequencies $2904 + 528 \times n_b$ MHz, $n_b = 1, 2, \ldots, 14$. The OFDM symbol is of duration $T_{FFT} = 242.42$ ns. After adding the cyclic prefix of length $T_{CP} = 60.61$ ns and the guard interval of length $T_{GI} = 9.47$ ns, the symbol duration becomes $T_{SYM} = 312.5$ ns. As in [Bat04], a convolutional encoder with a constraint length of 7 is used to generate different channel coding rates. The maximum transmitting power is -41.3 dBm/MHz, and the PER is maintained such that PER $< 8\%$ for a 1024-byte packet. The average noise power follows (9.6) with $N_F = 6.6$ dB, and the propagation loss factor is $v = 2$.

We consider a multiuser scenario in which each user is located at a distance of less than 4 m from the central base station. The performance is evaluated in multipath channel environments specified in the IEEE 802.15.3a channel modeling subcommittee report [Foe03b]. We employ channel models 1 and 2, which are based on channel measurements over the range 0 to 4 m. The simulated channels were constant during the transmission of each packet, and independent from one packet to another. In each simulation we averaged over a minimum of 50,000 channel realizations.

9.3.1 Subband Assignment and Power Allocation

In this subsection we present the average transmitting power and outage probability curves for UWB multiband OFDM systems. Here the outage probability is the probability that the rate requested cannot be supported under the constraints in (9.23). We compare the performance of the PCCA scheme with those of the current multiband scheme in the standards proposal [Bat04].

For Fig. 9.2, the number of users is fixed at $K = 3$, while each user is randomly located at the distance of 1 to 4 m from the base station. In Fig. 9.2(a) we illustrate the average transmitter power as a function of the data rates for the standard multiband scheme, fast suboptimal PCCA scheme, and optimal PCCA scheme obtained by full search. It is apparent that the suboptimal PCCA scheme can achieve almost the same performance as the optimal scheme. In addition, the PCCA scheme greatly reduces the average transmitter power compared to that in the standard proposal. The performance gain in terms of power reduction that is achieved by the PCCA scheme compared with the standard scheme can be computed as $(P_{standard} - P_{PCCA})/P_{standard}$, where $P_{standard}$ and P_{PCCA} denote the average powers for the standard and PCCA schemes, respectively. The results in Fig. 9.2(a) show that both fast suboptimal and optimal PCCA schemes can reduce about 60% of the average transmitter power at low rates (53.3 to 200 Mbps) and up to 35% at high rates (320 to 480 Mbps). Notice that the curves are not smooth because of the discrete nature of the problem. Figure 3(b) shows the outage probability versus the transmission rates. We can see that the PCCA scheme achieves a lower outage probability than that of the standard multiband scheme for any rates. For instance, at 110 Mbps, the outage probability of

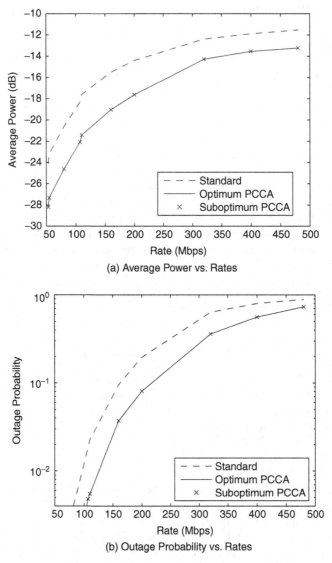

(a) Average Power vs. Rates

(b) Outage Probability vs. Rates

Figure 9.2 Performance of a three-user system with random location.

the PCCA scheme is 5.5×10^{-3}, whereas that of the standard multiband scheme is 2.3×10^{-2}.

We also consider a multiuser system with a different number of users, each located at a fixed position of about 4 m from the base station. Specifically, the distance between the kth user and the base station is specified as $d_k = 4 - 0.1(k - 1)$ for $k = 1, 2, \ldots, K$. In Fig. 9.3 we show the average transmitting power and outage probability as functions of the number of users for data rates of 55, 80, and 110 Mbps. In both figures we use the

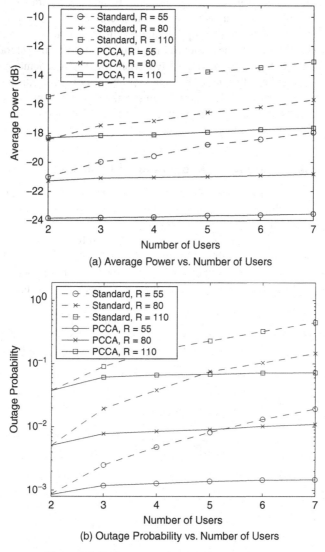

(a) Average Power vs. Number of Users

(b) Outage Probability vs. Number of Users

Figure 9.3 Performance of a multiple-user system.

standard multiband scheme and the PCCA scheme. We can observe from Fig. 9.3(a) that the transmitting power increases with the number of users. This results from the limited available subbands with good channel conditions. When the number of users is large, some users have to occupy those subbands that have the poorest channel conditions. Comparing the PCCA scheme with the standard multiband approach, we can see that the PCCA scheme achieves a lower transmitting power for all the rate requirements.

Figure 9.3(b) shows that the outage probability increases with the number of users. This is due to the fact that as the number of users increases, the system is more crowded and may not be feasible to support all these users at all times. Observe that at any rate, the performance of the standard multiband OFDM scheme degrades as the number of users increases. On the other hand, when the PCCA scheme is employed, the effect of the number of users to the outage probability is insignificant when the rates are not higher than 110 Mbps. As we can see, the PCCA scheme achieves lower outage probabilities than those of the standard scheme under all conditions.

9.3.2 Joint Rate Assignment and Resource Allocation

In this subsection we illustrate the performance of the joint rate assignment and resource allocation algorithm for a multiband OFDM system. We consider a multiuser system with a different number of users. Each user is randomly located 1 to 4 m from the base station. The user rates are also selected randomly from the set $\{200, 320, 400, 480\}$ Mbps, and the minimum rate requirement is $R_k^{\min} = 50$ Mbps $\forall k$ for the proportional fairness goal. The joint rate assignment and resource allocation algorithm presented in Section 9.2.5 is performed for each set of rates and channel conditions requested.

Figure 9.4 illustrates one realization of rate adaptation for a two-user system with three different goals. The shaded area represents the feasible range for R_1 and R_2 under current channel conditions. In this example the rates requested are $R_1 = 480$

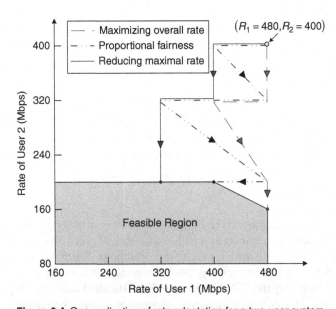

Figure 9.4 One realization of rate adaptation for a two-user system.

and $R_2 = 400$ Mbps, and both users are located about 4 m from the base station. We can observe from Fig. 9.4 that the reducing maximal rate approach has the lowest overall rate in every adaptation step. This is because the highest rate in the system can always be reduced. On the other hand, the maximizing overall rate approach tends to reduce the lower rate since most low rates have a smaller decreasing step size than that of high rates. Although the maximizing overall rate approach always yields superior system performance, it is unfair to those applications with low data rates. The proportional fairness goal provides a performance that is between the maximizing overall rate approach and the reducing maximal rate approach.

Figure 9.5 shows the average system performance versus the number of users. In Fig. 9.5(a) we present performance in terms of the average data rates of users. We can see that the average rates of all three approaches decrease when the number of users increases. This is due to the limited number of subbands under good channel conditions. As the number of users increases, some users need to occupy subbands with poor channel conditions, and hence their feasible rates tend to be lower than the rates requested. Comparing the performances of the three approaches, we can see that proportional fairness yields a slightly lower average rate than that of the maximizing overall rate approach, and both the proportional fairness and maximizing overall rate approaches achieve much higher rates than that of the reducing maximal rate approach.

In Fig. 9.5(b) we show the standard deviations of the user data rates for the three approaches. Here the standard deviation represents the fairness of allocation among users. We can observe that the standard deviation for every scheme increases with the number of users since the larger the number of users, the higher the variation of the rates. At any fixed number of users, the reducing maximal rate approach results in the lowest standard deviation, and its standard deviation increases slightly with the number of users. This is because the feasible rates obtained from the reducing maximal rate approach are close to each other. In contrast, the maximizing overall rate scheme can yield feasible rates of around 100 to 480 Mbps at the same time. Thus, its standard deviation increases much faster with the number of users. The standard deviation of the proportional fairness approach is between those of the other two schemes. So the proportional fairness approach is a trade-off between the maximal rate approach and the reducing maximal rate approach for both performance and fairness.

9.4 CHAPTER SUMMARY

Low power consumption is a key element in promoting UWB multiband OFDM technology as the solution for future indoor wireless communications. In this chapter we present an efficient power control channel allocation (PCCA) scheme for allocating subband and power among users in a UWB multiband OFDM system. The PCCA scheme aims to reduce power consumption without compromising performance. A general framework is provided to minimize the overall transmit power under practical implementation constraints. The problem formulated is NP hard; however, with the

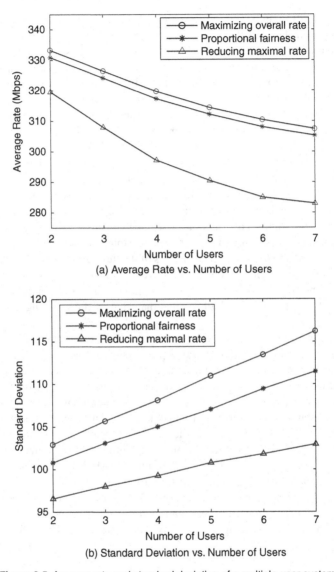

Figure 9.5 Average rate and standard deviation of a multiple-user system.

fast suboptimal algorithm, we can reduce the computational complexity to $O(K^2S)$, where K is the number of users and S is the number of subbands.

Simulation results show that the fast algorithm achieves performances comparable to those of the complex optimal full search algorithm and can save up to 61% of power consumption compared to the multiband OFDM scheme currently proposed in the IEEE 802.15.3a standard. Moreover, the fast algorithm can obtain feasible solutions adaptively when the initial system is not feasible for the rate requirements of users. Among three different system optimization goals used in the rate adaptation algorithm, the proportional fairness approach turns out to be a trade-off between the maximal rate approach and the reducing maximal rate approach for both performances and fairness.

10

COOPERATIVE UWB
MULTIBAND OFDM

Due to limitations on the transmitter power level, any UWB system faces significant design challenges to achieve the desired performance and coverage range. To date, there have been few proposals to improve the coverage of UWB systems. One approach is through the use of analog repeaters as used in conventional cellular systems: for example, a pulse position modulation UWB repeater was proposed in [Cho04]. Although analog repeaters are simple, they suffer from noise amplification, which has confined their application to specific scenarios. Another approach that has been considered is the employment of MIMO technology in UWB systems. As presented in Chapter 5, UWB-MIMO systems can efficiently exploit the available spatial and frequency diversities, and hence significantly improve the UWB performance and coverage range. Nevertheless, it might not be easy to have multiple antennas installed in UWB devices. One possible way to overcome this problem and to benefit from the performance enhancement introduced by MIMO systems is through the use of cooperative communications in UWB.

The research works in [Wor03, Wor04, Sen03a, Sen03b] have proved the significant potential of cooperative diversity in wireless networks. Current UWB technology, on the other hand, relies on a noncooperative transmission, in which the diversity can be obtained only from MIMO coding or information repetition at the transmitter [Bat04, Fen04, Sir06a]. Furthermore, many UWB devices are expected to be in home and office environments; most of these devices are not in the active mode simultaneously, but they can be utilized as relays to assist active devices. Additionally, due to the TDMA mechanism of the MAC and the network structure of the IEEE 802.15.3a WPAN standard [TG3a], the cooperative protocols can be adopted in UWB WPANs. These facts motivate the use of cooperative diversity in UWB systems as an alternative approach to improving UWB performance and coverage without the need for additional antennas or network infrastructures.

Ultra-Wideband Communications Systems: Multiband OFDM Approach, By W. Pam Siriwongpairat and K. J. Ray Liu
Copyright © 2008 John Wiley & Sons, Inc.

Adapted with permission from © 2006 *IEEE Wireless Communications and Networking Conference*, Vol. 4, Apr. 2006, pp. 1854–1859.

In this chapter we present performance enhancement of UWB systems with cooperative protocols [Sir06c]. The framework presented is based on a decode-and-forward cooperative protocol; however, other cooperative protocols can be used in a similar way. The symbol error rate (SER) performance analysis and optimum power allocation are provided for cooperative UWB multiband OFDM systems. To capture the unique multipath-clustering property of UWB channels [Foe03b], the SER performance is characterized in terms of the cluster and the ray arrival rates. Based on the established SER formulations, we determine optimum power allocations for cooperative UWB multiband OFDM systems with two different objectives: minimizing overall transmitted power and maximizing system coverage. When the subbands are not fully occupied, we improve the performance of cooperative UWB systems further by allowing the source to repeat its information on one subband, while the relay helps forward the source information on another subband. The improved cooperative UWB scheme is compatible with the current multiband OFDM standard proposal [Bat03], which allows multiuser transmission using different subbands. Both analytical and simulation results show that the cooperative UWB scheme achieves up to 43% power saving and up to 85% coverage extension compared with noncooperative UWB at the same data rate. By allowing the source and the relay to transmit simultaneously, the performance of cooperative UWB can be further improved, up to a 52% power saving and 100% coverage extension compared with the noncooperative scheme.

In Section 10.2 we describe system models of noncooperative and cooperative UWB systems employing multiband OFDM. In Section 10.3 we analyze the SER performance of the cooperative UWB multiband OFDM system. In Section 10.4 we study the optimum power allocation with the objectives of minimizing overall transmitter power and maximizing coverage. An improved cooperative UWB scheme is presented in Section 10.5. Simulation results are given in Section 10.6.

10.1 COOPERATIVE COMMUNICATIONS

Cooperative diversity has recently emerged as a promising alternative to combat fading in wireless channels. The basic idea is that users or nodes in a wireless network share their information and transmit cooperatively as a virtual antenna array, thus providing diversity without the requirement for additional antennas at each node.

Consider a cooperative strategy for a wireless network, which can be a mobile ad hoc network or a cellular network. Each user (or node) in the network can be a source node that sends information to its destination, or it can be a relay node that helps transmit the other user's information. Under the cooperative strategy in [Wor04], signal transmission involves two phases. In phase 1, each source sends information to its destination, and the information is also received by other users in the network. In phase 2, each relay may forward the source information to the destination or remain idle. In both phases, all users transmit signals through orthogonal channels using a time-division multiple access (TDMA), frequency-division multiple access (FDMA), or code-division multiple access (CDMA) scheme [Wor03, Wor04, Sen03a, Sen03b].

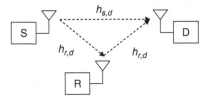

Figure 10.1 Simplified cooperative model.

Figure 10.1 illustrates a simplified cooperative model in which S, R, and D denote source, relay, and destination, respectively.

In [Wor04], the authors proposed various cooperative strategies, including fixed relaying, selection relaying, and incremental relaying schemes for single relay scenarios and analyzed their outage probability. The fixed relaying protocols include amplify-and-forward (AF) and decode-and-forward (DF) protocols. With the AF protocol, the relays simply amplify and forward the information, whereas with the DF protocol, the relays decode the information received and then forward the decoded symbols to the destination. In the selection relaying protocol, the relay decides whether to forward the information received from the source by applying a threshold test on the channel-state information measured between the source and the relay. With the incremental relaying protocol, limited feedback from the destination is employed in the form of automatic repeat request, and the relay forwards the source information only when the information is not captured successfully at the destination via direct transmission.

In [Wor03] the authors extended the DF cooperation in [Wor04] to the case of multiple relays, where they proposed distributed space–time coding. In [Sen03a, Sen03b], a similar concept, called user *cooperation diversity*, was proposed for CDMA systems in which orthogonal codes are used to mitigate multiple access interference. The work in [Sen03a, Sen03b] assumes full channel-state information at the cooperating nodes that utilize beamforming, while the protocols in [Wor04] assume no channel information at the transmitters since beamforming requires high-complexity radios and has not been demonstrated for the distributed case. In [Jan04], coded cooperation is proposed to achieve diversity by incorporating error control coding into cooperation. The scheme in [Jan04] does not use beamforming but assumes full channel-state information at the transmitter. In [Boy04] the authors introduced a concept of multihop diversity in which each relay combines the signals received from all previous transmissions. Later, in [Su05b], the authors provided SER performance analysis and optimum power allocation for decode-and-forward cooperation systems with two users. The SER performance analysis of a class of multinode cooperative protocols was presented in [Sad05].

10.2 SYSTEM MODEL

We consider the UWB multiband OFDM system [Bat04] presented in Chapter 4. The channel model is based on the S-V model described in Chapter 2.

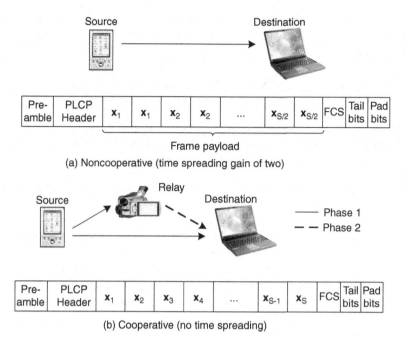

(a) Noncooperative (time spreading gain of two)

(b) Cooperative (no time spreading)

Figure 10.2 Noncooperative and cooperative UWB multiband OFDM systems with the same data rate.

10.2.1 Noncooperative UWB

In a noncooperative UWB multiband OFDM system, each source transmits information directly to its destination. We consider the case of time-domain spreading with a spreading factor of 2. In this scenario, the same information is transmitted repeatedly over two consecutive OFDM symbols, which can be sent on different subbands to gain the diversity from time spreading. Figure 10.2(a) depicts the frame structure for the multiband OFDM system with time spreading gain of 2. In Fig. 10.2, $\mathbf{x}_i$ ($1 \leq i \leq S$) denotes a vector of data symbols to be transmitted in each OFDM symbol, and S represents the number of OFDM symbols contained in the frame payload. With the choice of cyclic prefix length greater than the duration of the channel impulse response, OFDM allows each UWB subband to be divided into a set of orthogonal narrowband subcarriers. At the destination, the signal received at the nth subcarrier during the kth OFDM symbol duration can be modeled as

$$y_{s,d}^k(n) = \sqrt{P_k}\, H_{s,d}^k(n) x(n) + z_{s,d}^k(n), \tag{10.1}$$

where $x(n)$ denotes an information symbol to be transmitted at subcarrier n, $H_{s,d}^k(n)$ represents the frequency response of the channel from the source to the destination, and $z_{s,d}^k(n)$ is additive noise. The superscript index k, $k = 1$ and 2, is used to distinguish the signals in two consecutive OFDM symbols. In (10.1), P_k is the power transmitted at the source. As in the current multiband standard proposal [Bat03], we assume that the power P_k is equal for all subcarriers. Since time spreading is performed, $x(n)$ is the same in both OFDM symbols. The noise term $z_{s,d}^k(n)$ is modeled as a complex Gaussian random variable with zero mean and variance N_0. From (2.14) the channel frequency response is given by

$$H_{s,d}^k(n) = \sigma_{s,d}^2 \sum_{c=0}^{C} \sum_{l=0}^{L} \alpha_{s,d}^k(c, l) e^{-\mathrm{j}2\pi n\Delta f[T_{s,d}(c)+\tau_{s,d}(c,l)]}, \qquad (10.2)$$

where the subscript s,d indicates the channel link from the source to the destination. With ideal band hopping, we assume that the signal transmitted over different frequency bands undergones independent fading [i.e., $H_{s,d}^k(n)$ are independent for different k].

Note that with frequency-domain spreading, the same information can be transmitted in more than one subcarrier. For subsequent performance evaluation, we denote Φ_n as a set of subcarriers that carry the information $x(n)$. For instance, to minimize the correlation among the channel frequency response at different subcarriers, Φ_n can be given by [Su05a] $\Phi_n = \{n\}$ $(0 \leq n \leq N - 1)$ if $g_F = 1$ and $\Phi_n = \{n, n + N/2\}$ $(0 \leq n \leq N/2 - 1)$ if $g_F = 2$, where N is the total number of subcarriers and g_F represents the frequency spreading gain. Such frequency-domain spreading increases the frequency diversity and hence improves the performance of UWB systems with low data rates.

10.2.2 Cooperative UWB

We consider cooperative communications over a UWB multiband OFDM system with two users. This two-user cooperation will serve as a basic building block for future study of multiuser UWB systems. In a cooperative UWB system, each user can be a source node that sends its information to the destination, or it can be a relay node that helps transmit the other user's information. The cooperative strategy comprises two transmission phases. In phase 1, the source sends the information to its destination, and the information is received by other users at the same time. In phase 2, the source is silent, while the relay helps forward the source information. Suppose that the DF cooperative protocol is used; then the relay decodes the information received and forwards the decoded symbols to the destination. We consider the case when time-domain spreading is not performed at the source. In this scenario, the data frame that is transmitted from the source in phase 1 and from the relay in phase 2 can be depicted as in Fig. 10.2(b). Suppose that the noncooperative and cooperative UWB schemes have the same frequency spreading gain. Then we can see from Fig. 10.2(a) and (b) that the noncooperative UWB scheme with time spreading and the cooperative UWB scheme without time spreading achieve the same data rate.

In phase 1, the signal received at the destination is the same as (10.1) with $k = 1$, and the signal received at the relay can be written as

$$y_{s,r}(n) = \sqrt{P_1}\, H_{s,r}(n)x(n) + z_{s,r}(n), \qquad (10.3)$$

in which $H_{s,r}(n)$ is the channel frequency response from the source to the relay, and $z_{s,r}(n)$ is additive noise. In phase 2, the relay forwards the decoded symbol with power

P_2 to the destination only if the symbol is decoded correctly; otherwise, the relay does not send or remain idle. For mathematical tractability, we assume that the relay can judge whether the decoded information is correct.[1] The signal received at the destination in phase 2 can be specified as [Su05b]

$$y_{r,d}(n) = \sqrt{\tilde{P}_2} \, H_{r,d}(n) x(n) + z_{r,d}(n), \tag{10.4}$$

where $H_{r,d}(n)$ is the channel frequency response from the relay to the destination and $z_{r,d}(n)$ is additive noise. The transmitted power $\tilde{P}_2 = P_2$ if the relay correctly decodes symbol $x(n)$ transmitted from the source; otherwise, $\tilde{P}_2 = 0$ (i.e., the relay does not send or remains idle). The multipath channels of the source–relay and relay–destination links are also modeled according to the S-V model, with the total energy of the multipath components given by $\sigma_{s,r}^2$ and $\sigma_{r,d}^2$, respectively. The noise $z_{s,r}(n)$ and $z_{r,d}(n)$ are complex Gaussian distributed with zero mean and variance N_0. We assume that the channel-state information is known at the receiver but not at the transmitter. The channel coefficients are assumed to be independent for different transmitter–receiver links. As in noncooperative transmission, the information can be transmitted repeatedly in different subcarriers to obtain frequency diversity when the desired data rate is low. The destination employs a MRC [Sim00] to combine the information transmitted via different times (phase 1 and phase 2) or frequencies.

10.3 SER ANALYSIS FOR COOPERATIVE UWB

In this section we analyze the average SER performance of cooperative UWB multi-band OFDM systems with a DF protocol. Following the multiband standard proposal [Bat03], we focus on an analysis of UWB systems with M-PSK signals. The analysis for systems with M-QAM signals is similar.

10.3.1 Cooperative UWB

In this subsection we provide closed-form SER formulations for cooperative UWB systems. With knowledge of channel-state information, the destination detects the symbols transmitted by combining coherently the signals received from the source and the relay. Assume that each symbol transmitted has unit energy; then the instantaneous SNR of the MRC output can be written as [Sim00]

$$\eta = \frac{P_1}{N_0} \sum_{n \in \Phi_n} |H_{s,d}(n)|^2 + \frac{\tilde{P}_2}{N_0} \sum_{n \in \Phi_n} |H_{r,d}(n)|^2, \tag{10.5}$$

where Φ_n is the set of subcarriers that carry the information $x(n)$ as defined in Section 10.2. Suppose that the M-PSK modulation is used; then the conditional SER can be

---

[1] Practically, this can be done at the relay by applying a simple SNR threshold to the data received. Although it can lead to some error propagation, for practical ranges of operating SNR, the event of error propagation can be assumed negligible.

expressed as [Sim00]

$$P_e|_{\{H\}} = \Psi(\eta) \triangleq \frac{1}{\pi} \int_0^{\pi-\pi/M} \exp\left(-\frac{b\eta}{\sin^2\theta}\right) d\theta, \tag{10.6}$$

where $b = \sin^2(\pi/M)$. Recall thatthe relay forwards symbol $x(n)$ with power P_2 to the destination only if the symbol is decoded correctly. That is, $\tilde{P}_2 = P_2$ if the relay decodes correctly the symbol transmitted; otherwise, $\tilde{P}_2 = 0$. Assume that the relay has perfect knowledge of the channel gain coefficients $H_{s,r}(n)$, and the MRC is used to combine the information transmitted via different frequencies. Then the instantaneous SNR at the MRC output is given by $\eta_{s,r} = (P_1/N_0)\sum_{n\in\Phi_n}|H_{s,r}(n)|^2$, and the conditional probability of incorrect decoding at the relay is $\Psi(\eta_{s,r})$. Taking into account the two possible cases of $\tilde{P}_2$, the conditional SER in (10.6) can be reexpressed as

$$P_e|_{\{H\}} = \Psi(\eta)|_{\tilde{P}_2=0}\Psi(\eta_{s,r}) + \Psi(\eta)|_{\tilde{P}_2=P_2}[1 - \Psi(\eta_{s,r})]. \tag{10.7}$$

Substitute(10.5) into (10.7) and average over the channel realizations, resulting in the average SER:

$$\begin{aligned}
P_e &= \frac{1}{\pi^2} \int_0^{\pi-\pi/M} \mathcal{M}_{\eta_{s,d}}\left(\frac{b}{\sin^2\theta}\right) d\theta \int_0^{\pi-\pi/M} \mathcal{M}_{\eta_{s,r}}\left(\frac{b}{\sin^2\theta}\right) d\theta \\
&+ \frac{1}{\pi} \int_0^{\pi-\pi/M} \mathcal{M}_{\eta_{s,d}}\left(\frac{b}{\sin^2\theta}\right) \mathcal{M}_{\eta_{r,d}}\left(\frac{b}{\sin^2\theta}\right) d\theta \\
&\times \left[1 - \frac{1}{\pi} \int_0^{\pi-\pi/M} \mathcal{M}_{\eta_{s,r}}\left(\frac{b}{\sin^2\theta}\right) d\theta\right],
\end{aligned} \tag{10.8}$$

where $\mathcal{M}_\eta(s) = \mathrm{E}[\exp(-s\eta)]$ is the MGF of η [Sim00], $\eta_{s,d} = (P_1/N_0) \sum_{n\in\Phi_n}|H_{s,d}(n)|^2$, and $\eta_{r,d} = (P_2/N_0)\sum_{n\in\Phi_n}|H_{r,d}(n)|^2$. Note that the channel frequency responses, and hence the MGFs of $\eta_{s,d}$, $\eta_{s,r}$, and $\eta_{r,d}$, are in terms of multipath gain coefficients whose amplitudes are Rayleigh distributed, as well as the multipath delays T_c and $\tau_{c,l}$, which are based on the Poisson process. If the information is not jointly encoded across subcarriers (i.e., the frequency spreading gain is $g_F = 1$), $\mathcal{M}_{\eta_{x,y}}(s)$ can be determined as

$$\mathcal{M}_{\eta_{x,y}}(s) = \left(1 + \frac{sP_x\sigma_{x,y}^2}{N_0}\right)^{-1}, \tag{10.9}$$

where $P_x = P_1$ if x represents the source and $P_x = P_2$ if x represents the relay. If the data are jointly encoded across multiple subcarriers, it is difficult, if not impossible, to obtain closed-form formulations of the MGFs in (10.8). In this case we exploit an approximation approach in Chapter 9 which allows us to approximate the MGF of $\eta_{x,y}$ as

$$\mathcal{M}_{\eta_{x,y}}(s) \approx \prod_{n\in\Phi_n}\left(1 + \frac{sP_x\sigma_{x,y}^2\beta_n(\mathbf{R}_{x,y})}{N_0}\right)^{-1}, \tag{10.10}$$

where $\beta_n(\mathbf{R}_{x,y})$ denotes the eigenvalues of a matrix $\mathbf{R}_{x,y}$, and $\mathbf{R}_{x,y}$ is a correlation matrix each diagonal component of which is 1, and the (i, j)th $(i \neq j)$ component is given by

$$
\mathbf{R}_{x,y}(i, j) = \Omega_{x,y}(0, 0) \frac{\Lambda_{x,y} + \Gamma_{x,y}^{-1} + \mathbf{j}2\pi(n_i - n_j)\Delta f}{\Gamma_{x,y}^{-1} + \mathbf{j}2\pi(n_i - n_j)\Delta f}
$$

$$
\times \frac{\lambda_{x,y} + \gamma_{x,y}^{-1} + \mathbf{j}2\pi(n_i - n_j)\Delta f}{\gamma_{x,y}^{-1} + \mathbf{j}2\pi(n_i - n_j)\Delta f}, \tag{10.11}
$$

in which n_i denotes the ith element in the set Φ_n. By substituting the MGFs in (10.9) and (10.10) into (10.8), we can express the SER of a cooperative UWB system as

$$
P_e \approx F\left(\prod_{n \in \Phi_n}\left(1 + \frac{bP_1\sigma_{s,d}^2\beta_n(\mathbf{R}_{s,d})}{N_0 \sin^2 \theta}\right)\right) F\left(\prod_{n \in \Phi_n}\left(1 + \frac{bP_1\sigma_{s,r}^2\beta_n(\mathbf{R}_{s,r})}{N_0 \sin^2 \theta}\right)\right)
$$

$$
+ F\left(\prod_{n \in \Phi_n}\left(1 + \frac{bP_1\sigma_{s,d}^2\beta_n(\mathbf{R}_{s,d})}{N_0 \sin^2 \theta}\right)\left(1 + \frac{bP_2\sigma_{r,d}^2\beta_n(\mathbf{R}_{r,d})}{N_0 \sin^2 \theta}\right)\right)
$$

$$
\times \left[1 - F\left(\prod_{n \in \Phi_n}\left(1 + \frac{bP_1\sigma_{s,r}^2\beta_n(\mathbf{R}_{s,r})}{N_0 \sin^2 \theta}\right)\right)\right], \tag{10.12}
$$

where

$$
F(x(\theta)) = \frac{1}{\pi}\int_0^{\pi - \pi/M} \frac{1}{x(\theta)}\, d\theta. \tag{10.13}
$$

Note that the average SER in (10.12) is exact if $\Phi_n = \{n\}$ (i.e., the frequency spreading gain is $g_F = 1$).

In (10.12) we provide an SER formulation for general cooperative UWB systems. Such an SER formulation involves integrations, so it is difficult to develop insightful understanding of the UWB system performance. To get more insight, we provide in what follows SER approximations that involve no integrations. We focus on the SER performance of two special cases that have been considered in the multiband standard proposal [Bat03].

1. If the frequency spreading gain is $g_F = 1$, the average SER can be expressed as

$$
P_e = F\left(1 + \frac{bP_1\sigma_{s,d}^2}{N_0 \sin^2 \theta}\right) F\left(1 + \frac{bP_1\sigma_{s,r}^2}{N_0 \sin^2 \theta}\right)
$$

$$
+ F\left(\left(1 + \frac{bP_1\sigma_{s,d}^2}{N_0 \sin^2 \theta}\right)\left(1 + \frac{bP_2\sigma_{r,d}^2}{N_0 \sin^2 \theta}\right)\right)\left[1 - F\left(1 + \frac{bP_1\sigma_{s,r}^2}{N_0 \sin^2 \theta}\right)\right], \tag{10.14}
$$

which is the same as that of a cooperative narrowband system in a Rayleigh fading environment. It has been shown in [Su05b] that when all channel links are available (i.e., $\sigma_{s,d}^2 \neq 0$, $\sigma_{s,r}^2 \neq 0$, and $\sigma_{r,d}^2 \neq 0$), the SER (10.14) can be upper bounded by

$$P_e \leq \frac{A_1^2}{b^2 \rho_1^2 \sigma_{s,d}^2 \sigma_{s,r}^2} + \frac{A_2}{b^2 \rho_1 \rho_2 \sigma_{s,d}^2 \sigma_{r,d}^2}, \tag{10.15}$$

where $\rho_i = P_i/N_0$ and

$$A_i = \frac{1}{\pi} \int_0^{\pi - \pi/M} \sin^{2i} \theta \, d\theta. \tag{10.16}$$

Specifically, we have [Su05b]

$$A_1 = \frac{M-1}{2M} + \frac{1}{4\pi} \sin \frac{2\pi}{M}$$

and

$$A_2 = \frac{3(M-1)}{8M} + \frac{1}{4\pi} \sin \frac{2\pi}{M} - \frac{1}{32\pi} \sin \frac{4\pi}{M}.$$

The upper bound in (10.15) is loose at low SNR, but it is tight at high SNR [Su05b]. However, UWB systems may operate at low SNR due to the limitation on the transmitter power level. In what follows we provide an SER approximation that is close to the exact SER for every SNR and does not involve integrations. Observe that all the integrands on the right-hand side of (10.14) can be written as $F((p(\sin^2 \theta) + c)/\sin^{2i} \theta)$, where i is a positive integer, c is a constant that does not depend on θ, and $p(x)$ denotes a polynomial function of x. By bounding $p(\sin^2 \theta)$ with $p(1)$ and removing the negative term in (10.14), the SER can be approximated by

$$P_e \approx \frac{A_1^2}{1 + b\rho_1 \left(\sigma_{s,d}^2 + \sigma_{s,r}^2\right) + b^2 \rho_1^2 \sigma_{s,d}^2 \sigma_{s,r}^2}$$
$$+ \frac{A_2}{1 + b\left(\rho_1 \sigma_{s,d}^2 + \rho_2 \sigma_{r,d}^2\right) + b^2 \rho_1 \rho_2 \sigma_{s,d}^2 \sigma_{r,d}^2}. \tag{10.17}$$

2. If frequency spreading gain is $g_F = 2$, the eigenvalues of the correlation matrix $\mathbf{R}_{x,y}$ are $1 + B_{x,y}$ and $1 - B_{x,y}$, where

$$B_{x,y} = \Omega_{x,y}(0,0) \frac{\left[(\Lambda_{x,y} + \Gamma_{x,y}^{-1})^2 + q\right]^{1/2} \left[(\lambda_{x,y} + \gamma_{x,y}^{-1})^2 + q\right]^{1/2}}{\left[(\Gamma_{x,y}^{-1})^2 + q\right]^{1/2} \left[(\gamma_{x,y}^{-1})^2 + q\right]^{1/2}}, \tag{10.18}$$

in which $q = (2\pi \mu \Delta f)^2$ and μ denotes the subcarrier separation. Substituting the eigenvalues of correlation matrices $\mathbf{R}_{s,d}$, $\mathbf{R}_{s,r}$, and $\mathbf{R}_{r,d}$ into (10.12), we can simplify the approximate SER to

$$P_e \approx F(V_{s,d})F(V_{s,r}) + F(V_{s,d}V_{r,d})[1 - F(V_{s,r})], \qquad (10.19)$$

where

$$V_{x,y} = 1 + \frac{bP_x\sigma_{x,y}^2}{N_0 \sin^2\theta}\left(1 + \frac{bP_x\sigma_{s,d}^2(1 - B_{x,y}^2)}{N_0 \sin^2\theta}\right).$$

Following the same approximation approach as in [Su05b], we obtain an approximate SER at high SNR as

$$P_e \approx \frac{A_2^2}{b^4\rho_1^4\sigma_{s,d}^4\sigma_{s,r}^4\left(1 - B_{s,d}^2\right)\left(1 - B_{s,r}^2\right)}$$
$$+ \frac{A_4}{b^4\rho_1^2\rho_2^2\sigma_{s,d}^4\sigma_{r,d}^4\left(1 - B_{s,d}^2\right)\left(1 - B_{r,d}^2\right)}. \qquad (10.20)$$

Similar to the case of no frequency spreading, a tighter approximate SER can be obtained by replacing $p(\sin^2\theta)$ with $p(1)$. The resulting SER can be expressed as

$$P_e \approx \frac{1}{1 + b\rho_1\sigma_{s,d}^2 + b^2\rho_1^2\sigma_{s,d}^4\left(1 - B_{s,d}^2\right)}$$
$$\times\left(\frac{A_2^2}{1 + b\rho_1\sigma_{s,r}^2 + b^2\rho_1^2\sigma_{s,r}^4\left(1 - B_{s,r}^2\right)}\right.$$
$$\left.+ \frac{A_4}{1 + b\rho_2\sigma_{r,d}^2 + b^2\rho_2^2\sigma_{r,d}^4\left(1 - B_{r,d}^2\right)}\right). \qquad (10.21)$$

In Fig. 10.3 we compare the SER approximations above with SER simulation curves for a cooperative UWB system with frequency spreading gains of 1 and 2. The simulated multiband OFDM system has $N = 128$ subcarriers, the subband bandwidth is 528 MHz, and the channel model parameters follow those for CM1 [Foe03b]. For a frequency spreading gain $g_F = 2$, the subcarrier separation is chosen as $\mu = N/2 = 64$. For fair comparison, we plot average SER curves as functions of P/N_0. For a frequency spreading gain g_F of 1, the theoretical calculation (10.14) matches the simulation curve. With a frequency spreading gain of 2, the SER approximation (10.19) is also close to the simulation curve, except for some difference at a low SNR which is due to the approximation of the Poisson behavior of the multipath components. The SER approximations (10.15) and (10.20) are loose at low SNR but are tight at high SNR, as expected. Moreover, the SER approximations (10.17) and (10.21) are close to the simulation curves for the entire SNR range.

It is worth noting that the SER analysis provided in this section includes a two-hop relay communication scenario as a special case. Specifically, the performance of the two-hop relay system can be obtained from (10.12) by replacing $\sigma_{s,d}^2$ with 0.

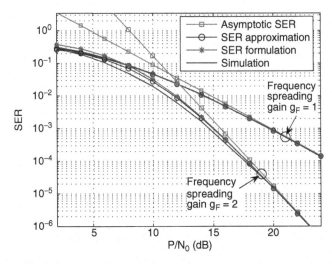

Figure 10.3 Comparison of the SER formulations and the simulation result for a cooperative UWB system. We assume that $\sigma_{s,d}^2 = \sigma_{s,r}^2 = \sigma_{r,d}^2 = 1$, and $P_1 = P_2 = P/2$.

The resulting SER of a two-hop relay cooperative UWB system can be tightly upper bounded at high SNR as

$$P_e \leq \frac{A_1 N_0}{b} \left(\frac{1}{P_1 \sigma_{s,r}^2} + \frac{1}{P_2 \sigma_{r,d}^2} \right) \qquad \text{if } g_F = 1 \quad (10.22)$$

$$P_e \approx \frac{A_2^2 N_0^2}{b^2} \left(\frac{1}{P_1^2 \sigma_{s,r}^4 (1 - B_{s,r}^2)} + \frac{1}{P_2^2 \sigma_{r,d}^4 (1 - B_{r,d}^2)} \right) \quad \text{if } g_F = 2. \quad (10.23)$$

10.3.2 Comparison of Cooperative and Noncooperative UWB

In this subsection we compare the performance of cooperative and noncooperative UWB multiband OFDM systems that have the same transmission data rate. Consider a noncooperative UWB system with a time spreading gain of 2, as described in Section 10.2.1. With an assumption of ideal band hopping, the average SER can be given by

$$P_e \approx F \left(\prod_{n \in \Phi_n} \left(1 + \frac{b P_1 \sigma_{s,d}^2 \beta_n(\mathbf{R}_{s,d})}{N_0 \sin^2 \theta} \right) \prod_{n \in \Phi_n} \left(1 + \frac{b P_2 \sigma_{s,d}^2 \beta_n(\mathbf{R}_{s,d})}{N_0 \sin^2 \theta} \right) \right). \quad (10.24)$$

In a noncooperative UWB system, power is generally applied to the source equally in two time slots [Bat04]. By letting $P_1 = P_2 = P/2$ and removing all the 1's in (10.24), the SER of noncooperative UWB systems can be expressed as

$$
P_e \approx \begin{cases} \left(\dfrac{b\sigma_{s,d}^2}{2\sqrt{A_2}} \dfrac{P}{N_0} \right)^{-2} & \text{if } g_F = 1; \qquad (10.25) \\[4mm] \left(\dfrac{b\sigma_{s,d}^2 \sqrt{1 - B_{s,d}^2}}{2A_4^{\frac{1}{4}}} \dfrac{P}{N_0} \right)^{-4} & \text{if } g_F = 2. \qquad (10.26) \end{cases}
$$

The results above indicate that the diversity order of a noncooperative UWB system with time spreading is twice the frequency spreading gain ($2g_F$), as expected. Moreover, the coding gain is $G_{NC} = b\sigma_{s,d}^2/2\sqrt{A_2}$ if the frequency spreading gain $g_F = 1$ and $G_{NC} = b\sigma_{s,d}^2\sqrt{1 - B_{s,d}^2}/2A_4^{1/4}$ if $g_F = 2$.

In cooperative systems we do not really have the notion of coding since the information is not encoded jointly at the source. However, combining the signals transmitted from the direct and relay links also results in system performance of a form $P_e = (G_{DF}P/N_0)^{-\Delta}$, where Δ is the diversity gain and G_{DF} represents the overall cooperative gain of cooperative UWB systems. Let us denote $r = P_1/P$ as the power ratio of the power P_1 transmitted at the source over the total power P. According to (10.15) and (10.20), the approximate SER of a cooperative UWB system can be expressed as

$$
P_e \approx \begin{cases} \left(\dfrac{b\sigma_{s,d}\sigma_{s,r}\sigma_{r,d}r}{\sqrt{A_1^2\sigma_{r,d}^2 + A_2\sigma_{s,r}^2 r/(1-r)}} \dfrac{P}{N_0} \right)^{-2} & \text{if } g_F = 1; \\[2mm] & \qquad (10.27) \\[4mm] \left(\dfrac{b\sigma_{s,d}\sigma_{s,r}\sigma_{r,d}r[(1 - B_{s,d}^2)(1 - B_{s,r}^2)(1 - B_{r,d}^2)]^{\frac{1}{4}}}{[A_2^2\sigma_{r,d}^2(1 - B_{r,d}^2) + A_4\sigma_{s,r}^2(1 - B_{s,r}^2)r^2/(1-r)^2]^{\frac{1}{4}}} \dfrac{P}{N_0} \right)^{-4} & \text{if } g_F = 2. \\[2mm] & \qquad (10.28) \end{cases}
$$

We can see that the cooperative UWB systems also achieve the diversity gain of twice the frequency spreading gain. However, the cooperative gain depends not only on the channel quality of the source–destination link, but also on the channel qualities of the source–relay link as well as of the relay–destination link. Since both noncooperative and cooperative UWB systems achieve the same diversity order, it is interesting to compare the coding and cooperative gains. We define the ratio between these two gains as $\xi = G_{DF}/G_{NC}$. From the SER expressions in (10.25)–(10.28), we have

$$
\xi = \begin{cases} \dfrac{2\sigma_{s,r}\sigma_{r,d}r}{\sigma_{s,d}} \left(\dfrac{A_1^2}{A_2}\sigma_{r,d}^2 + \dfrac{r}{1-r}\sigma_{s,r}^2 \right)^{-1/2} & \text{if } g_F = 1; \\[2mm] & \qquad (10.29) \\[4mm] \dfrac{2\sigma_{s,r}\sigma_{r,d}r}{\sigma_{s,d}} \left(\dfrac{A_2^2}{A_4}\dfrac{(1 - B_{s,d}^2)}{(1 - B_{s,r}^2)}\sigma_{r,d}^2 + \dfrac{r^2}{(1-r)^2}\dfrac{(1 - B_{s,d}^2)}{(1 - B_{r,d}^2)}\sigma_{s,r}^2 \right)^{-1/4} & \text{if } g_F = 2. \\[2mm] & \qquad (10.30) \end{cases}
$$

Note that if all the channel links have the same qualities (e.g., $\sigma_{s,d}^2 = \sigma_{s,r}^2 = \sigma_{r,d}^2 = 1$), then $\xi < 1$ for any value of the power ratio $0 < r < 1$, which implies that noncooperative transmission is preferable. The reason for this is that signals from the source and from the relay are sent through the links with equal qualities. However, the source is the most reliable node since it has the original copy of the signals, while the relay may not be able to acquire the original signal due to the noisy channel between the source and the relay. As a result, noncooperative systems all of whose signals come from the source yield perform better than a cooperative system in which some of the signals come from the relay. On the other hand, when the link between the source and the relay or that between the relay and the destination is of better quality than the source–destination link (e.g., when the relay is located between the source and the destination), the DF cooperation gain, G_{DF}, could be greater than the coding gain, G_{NC}, depending on the power ratio r and the channel qualities. In Section 10.4 we determine the power ratio and the relay location that lead to the optimum performance of cooperative UWB systems.

10.4 OPTIMUM POWER ALLOCATION FOR COOPERATIVE UWB

In this section we provide the optimum power allocation for a cooperative UWB multiband OFDM system with two different objectives: minimizing overall power transmitted and maximizing coverage. First, we formulate a problem to minimize the overall power transmitted under the constraints on performance requirement and power spectral density limitation. The optimum power allocation is determined based on the tight SER approximations in Section 10.3. Then we determine an optimum power allocation such that the coverage of UWB system is maximized.

10.4.1 Power Minimization Using Cooperative Communications

In this subsection we determine optimum power allocation based on the SER formulations derived in Section 10.3. Our objective is to minimize the overall power transmitted under the constraint on SER performance and the transmitter power level. For notational convenience, let us define $\mathbf{P} = [P_1 \ P_2]^T$ as a power allocation vector. Now we can formulate the optimization problem as

$$\min_{\mathbf{P}} \ P = \sum_i P_i \tag{10.31}$$

$$\text{s.t.} \begin{cases} \text{Performance: } P_e \leq \varepsilon \\ \text{Power: } P_i \leq P_{\max}, \ \forall i, \end{cases}$$

where ε denotes the SER required and $P_{\max}$ is the maximum power transmitted for each subcarrier. The first constraint in (10.31) is to ensure the performance requirement. The average SER P_e follows the SER formulation in (10.8). The second constraint is related to the limitation on the transmitter power level.

For simplicity and for better understanding of system performance, let us consider at first the formulated problem in (10.31) without the maximum power constraint.

Applying the Lagrange multiplier method, the optimum power allocation can be obtained by solving $1 + \zeta \partial P_e/\partial P_1 = 0$, $1 + \zeta \partial P_e/\partial P_2 = 0$, and $P_e - \varepsilon = 0$, where ζ represents the Lagrange multiplier. In Section 10.3 we provide theoretical SER approximations that are close to the simulated SER. Based on such SER approximations, we can determine the optimum power allocation as follows. According to the tight SER approximations (10.17) and (10.21), the optimum power allocation for a cooperative UWB system can be obtained by solving the following equations:

$$\frac{A_i^2}{f_{s,r}}\left(\frac{1}{f_{s,r}}\frac{\partial f_{s,r}}{\partial P_1} + \frac{1}{f_{s,d}}\frac{\partial f_{s,d}}{\partial P_1}\right) - \frac{A_{2i}}{f_{r,d}}\left(\frac{1}{f_{s,d}}\frac{\partial f_{s,d}}{\partial P_1} - \frac{1}{f_{r,d}}\frac{\partial f_{r,d}}{\partial P_2}\right) = 0$$

$$\frac{1}{f_{s,d}}\left(\frac{A_i^2}{f_{s,r}} + \frac{A_{2i}}{f_{r,d}}\right) - \varepsilon = 0, \quad (10.32)$$

where i denotes the frequency spreading gain and A_i is specified in (10.16). If the frequency spreading gain is $g_F = 1$, $f_{s,d} = 1 + b\rho_1\sigma_{s,d}^2$, $f_{s,r} = 1 + b\rho_1\sigma_{s,r}^2$, and $f_{r,d} = 1 + b\rho_2\sigma_{r,d}^2$. If the frequency spreading gain is $g_F = 2$, $f_{s,d} = 1 + b\rho_1\sigma_{s,d}^2 + b^2\rho_1^2\sigma_{s,d}^4(1 - B_{s,d}^2)$, $f_{s,r} = 1 + b\rho_1\sigma_{s,r}^2 + b^2\rho_1^2\sigma_{s,r}^4(1 - B_{s,r}^2)$, and $f_{r,d} = 1 + b\rho_2\sigma_{r,d}^2 + b^2\rho_2^2\sigma_{r,d}^4(1 - B_{r,d}^2)$. At high enough SNR, the asymptotic optimum power allocation can be obtained from the tight SER upper bound (10.15) in case of $g_F = 1$ and from the SER approximation (10.20) in case of $g_F = 2$. According to the SER upper bound in (10.15), the asymptotic optimum power allocation for a cooperative UWB system with $g_F = 1$ can be determined as

$$P_1 = rP \quad \text{and} \quad P_2 = (1 - r)P, \quad (10.33)$$

where

$$P = \frac{N_0}{br\sigma_{s,d}\sigma_{s,r}\sigma_{r,d}}\left(\frac{A_2 r\sigma_{s,r}^2 + A_1^2(1 - r)\sigma_{r,d}^2}{\varepsilon(1 - r)}\right)^{1/2} \quad (10.34)$$

$$r = \frac{\sigma_{s,r} + \sqrt{\sigma_{s,r}^2 + (8A_1^2/A_2)\sigma_{r,d}^2}}{3\sigma_{s,r} + \sqrt{\sigma_{s,r}^2 + (8A_1^2/A_2)\sigma_{r,d}^2}}. \quad (10.35)$$

Based on the SER approximation (10.20), the asymptotic optimum power allocation for a system with frequency spreading gain $g_F = 2$ can be written in the same form as (10.33), with

$$P = \frac{N_0}{br\sigma_{s,d}\sigma_{s,r}\sigma_{r,d}}\left(\frac{A_4 r^2\sigma_{s,r}^4(1 - B_{s,r}^2) + A_2^2(1 - r)^2\sigma_{r,d}^4(1 - B_{r,d}^2)}{\varepsilon(1 - r)^2(1 - B_{s,d}^2)(1 - B_{s,r}^2)(1 - B_{r,d}^2)}\right)^{1/4}, \quad (10.36)$$

and r being the solution to the equation $(2c_{s,r} + c_{r,d})r^3 - (c_{s,r} + 3c_{r,d})r^2 + 3c_{r,d}r - c_{r,d} = 0$, where $c_{s,r} = A_4\sigma_{s,r}^4(1 - B_{s,r}^2)$ and $c_{r,d} = 2A_2^2\sigma_{r,d}^4(1 - B_{r,d}^2)$ are constants that depend on the average channel quality of the source–relay and relay–destination links, respectively. By solving the polynomial equation, we arrive after some manipulation at

$$r = \frac{4^{1/3}c^2 + 2(c_{s,r} + 3c_{r,d})c + 4^{2/3}\left(c_{s,r}^2 - 12c_{s,r}c_{r,d}\right)}{6(2c_{s,r} + c_{r,d})c}, \tag{10.37}$$

in which

$$c = \left(72c_{s,r}c_{r,d} + 2c_{s,r}^2 - 27c_{r,d}^2 + 3(2c_{s,r} + c_{r,d})\sqrt{3\left(4c_{s,r}c_{r,d} + 27c_{r,d}^2\right)}\right)^{1/3}.$$

The results in (10.35) and (10.37) reveal that the asymptotic power allocation of cooperative UWB systems with any frequency spreading gain does not rely on the channel link between the source and the destination. It depends only on the channel link between the source and the relay and the channel link between the relay and the destination. If the link quality between the source and the relay is the same as that between the relay and the destination, the power ratio is simplified to

$$r = \begin{cases} \dfrac{1 + \sqrt{1 + \left(8A_1^2/A_2\right)}}{3 + \sqrt{1 + \left(8A_1^2/A_2\right)}} & \text{if } g_F = 1; \tag{10.38} \\[2ex] \dfrac{c^2 + \left(A_4 + 6A_2^2\right)c - 24A_2^2A_4 + A_4^2}{6\left(A_4 + A_2^2\right)c} & \text{if } g_F = 2, \tag{10.39} \end{cases}$$

where

$$c = \left[A_4\left(18A_2^2\left(4A_4 - 3A_2^2\right) + A_4^2 + 6A_2\left(A_4 + A_2^2\right)\sqrt{3\left(2A_4 + 27A_2^2\right)}\right)\right]^{1/3},$$

and the A_i depend on specific modulation signals. If QPSK modulation is used, $r = 0.6207$ when the frequency spreading gain $g_F = 1$ and $r = 0.5925$ when $g_F = 2$. Observe from (10.38) and (10.39) that when the source–relay and relay–destination links are of the same quality, the asymptotic power ratio does not depend on the clustering property of UWB channels, regardless of the frequency spreading gain. In general, this is not the case, especially when frequency spreading is performed. As we can see from (10.36) and (10.37), the optimum power allocation for UWB system with a frequency spreading gain of 2 generally depends on both the channel gains and the multipath clustering property of UWB channels.

Table 10.1 provides comparisons between the optimum power allocation obtained via exhaustive search to minimize the SER formulation in (10.12), the one obtained by solving (10.32), and the one provided by the closed-form expressions in (10.34) and (10.36). The SER required performance is set at 5×10^{-2}. We consider the DF cooperative system under two different scenarios: $\sigma_{s,d}^2 = \sigma_{r,d}^2 = 1$ and $\sigma_{s,r}^2 = 10$ as well as $\sigma_{s,d}^2 = \sigma_{s,r}^2 = 1$ and $\sigma_{r,d}^2 = 10$. Channel model parameters of each channel link are based on CM1. We can see that the optimum power allocations obtained by solving (10.32) and by closed-form expressions in (10.34) and (10.36) agree with that obtained via exhaustive search for all scenarios considered. Furthermore, Table 10.1 illustrates that the optimum power allocation does not depend strongly on the spreading gain, but it relies mostly on the channel link quality. If the link quality

TABLE 10.1 Comparisons Between Optimum Power Allocation Obtained Via Exhaustive Search and Analytical Results

Multipath Energy							r		
$\sigma_{s,r}^2$	$\sigma_{s,r}^2$	$\sigma_{r,d}^2$	Gain, g_F	Search	From (10.32)	From (10.34) and (10.36)
1	10	1	1	0.5321	0.5356	0.5247
1	10	1	2	0.5072	0.5095	0.5023
1	1	10	1	0.7873	0.7772	0.7968
1	1	10	2	0.8082	0.7882	0.8316

between the source and the relay is much better than that between the relay and the destination, the power should be allocated equally at the source and the relay. If the source–relay link has much less quality than the relay–destination link, more power is allocated at the source. This is inconsistent with the results in [Su05b], in which it was shown that for a cooperative system to achieve a performance diversity of 2, the source–relay and relay–destination links should be balanced.

In the sequel we compare the total transmitted power used in noncooperative and cooperative systems to achieve the same SER performance. According to the SER expressions in Section 10.3.2, the ratio between the power of cooperative and noncooperative UWB systems with the same spreading gain can be expressed as

$$\frac{P_{DF}}{P_{NC}} = \frac{N_0 P_e^{-1/\Delta} G_{DF}^{-1}}{N_0 P_e^{-1/\Delta} G_{NC}^{-1}} = \frac{G_{NC}}{G_{DF}} = \frac{1}{\xi}. \tag{10.40}$$

Substituting (10.29) and (10.35) into (10.40), the ratio P_{DF}/P_{NC} for the UWB systems with frequency spreading gain $g_F = 1$ is given by

$$\frac{P_{DF}}{P_{NC}} = \frac{\sigma_{s,d}(3 + K_1)}{2(1 + K_1)} \left(\frac{A_1^2}{A_2 \sigma_{s,r}^2} + \frac{1 + K_1}{2\sigma_{r,d}^2} \right)^{1/2}, \tag{10.41}$$

where $K_1 = \sqrt{1 + 8A_1^2 \sigma_{r,d}^2 / A_2 \sigma_{s,r}^2}$. For systems with frequency spreading gain $g_F = 2$, the ratio P_{DF}/P_{NC} can be calculated from (10.30) and (10.37) as

$$\frac{P_{DF}}{P_{NC}} = \frac{6\sigma_{s,d}c(2c_{s,r} + c_{r,d})}{2\sigma_{s,r}\sigma_{r,d}(K_2 + 2c(c_{s,r} + 3c_{r,d}))} \left(\frac{A_2^2(1 - B_{s,d}^2)\sigma_{r,d}^2}{A_4(1 - B_{s,r}^2)} \right.$$

$$\left. + \frac{(K_2 + 2c(c_{s,r} + 3c_{r,d}))^2(1 - B_{s,d}^2)\sigma_{s,r}^2}{(K_2 - 10c_{s,r}c)^2(1 - B_{r,d}^2)} \right)^{1/4}, \tag{10.42}$$

where $K_2 = 4^{1/3}c^2 + 4^{2/3}(c_{s,r}^2 - 12c_{s,r}c_{r,d})$. Tables 10.2 demonstrates the ratios P_{DF}/P_{NC} for UWB systems with different channel qualities. The channel model parameters are the same for every link. In this scenario, (10.41) and (10.42) disclose that the ratio P_{DF}/P_{NC} does not depend on the clustering property of UWB channels. If all the channel links are of the same quality, noncooperative transmission requires

TABLE 10.2 Power Ratio of Cooperative and Noncooperative UWB Multiband OFDM Systems

Multipath Energy			P_{DF}/P_{NC}	
$\sigma_{s,d}^2$	$\sigma_{s,r}^2$	$\sigma_{r,d}^2$	$g_F = 1$	$g_F = 2$
1	1	1	1.7189	1.0709
1	10	1	0.5287	0.5689
1	1	10	0.2132	0.5545

less transmitted power than does cooperative transmission. However, if the channel quality of either source–relay link or relay–destination link is very good, cooperative transmission significantly reduces the power transmitted. As shown in Table 10.2, the cooperative scheme with a high-quality link between source and relay yields about a 50% power saving compared to the noncooperative scheme. For a high-quality link between the relay and the destination, the power transmitted using a cooperative scheme can be reduced by up to 78% over a noncooperative scheme.

We have determined the optimum power allocation for a cooperative UWB multiband OFDM system without taking into consideration limitations on the transmitter power level. With the maximum power limitation, it is difficult to obtain a closed-form solution to the problem in (10.31). In this case we provide a solution as follows. Let P_1 and P_2 be the power transmitted obtained by solving (10.31) without the maximum power constraint, and let $\hat{P}_1$ and $\hat{P}_2$ denote our solution.

- If $\min\{P_1, P_2\} > P_{max}$, there is no feasible solution to (10.31).
- Else, if $\max\{P_1, P_2\} \leq P_{max}$, $\hat{P}_1 = P_1$ and $\hat{P}_2 = P_2$.
- Otherwise:
  (i) Let $j = \text{argmax}_i\{P_1, P_2\}$ and $j' = \text{argmin}_i\{P_1, P_2\}$.
  (ii) Set $P_j = P_{max}$ and find $P_{j'}$ such that the SER performance desired is satisfied [i.e., $P_{j'}$ is obtained by solving $P_e - \varepsilon = 0$, where P_e is as expressed in (10.22) or (10.23), with P_j replaced by P_{max}].
  (iii) If the $P_{j'}$ obtained $\leq P_{max}$, then $\hat{P}_j = P_{max}$ and $\hat{P}_{j'} = P_{j'}$; otherwise, there is no feasible solution to (10.31).

The case of no feasible solution to (10.31) indicates that the UWB system under current channel conditions cannot satisfy the performance requirement even by exploiting the cooperative diversity. In this scenario, an additional subband can be utilized to further increase the diversity gain and improve system performance as discussed in Section 10.5.

10.4.2 Coverage Enhancement Using Cooperative Communications

Coverage of a UWB system can be specified by the maximum distance between the source and the destination at which the system is able to offer transmission with an

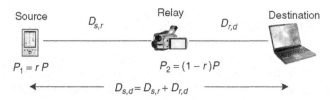

Figure 10.4 Coverage enhancement using cooperative UWB multiband OFDM.

error probability below the threshold value desired. In this subsection we determine the optimum power allocation and the relay location that will maximize coverage of a cooperative UWB multiband OFDM system.

We take into account the effect of the geometry on the channel link qualities by assuming that the total multipath energy between any two nodes is proportional to the distance between them. Particularly, the total multipath energy $\sigma_{x,y}^2$ is modeled by [Pro01]

$$\sigma_{x,y}^2 = \kappa D_{x,y}^{-\nu}, \tag{10.43}$$

where κ is a constant whose value depends on the propagation environment, ν is the propagation loss factor, and $D_{x,y}$ represents the distance between nodes x and y. Given a fixed total transmitted power P, we aim to find the optimum power allocation $r = P_1/P$ such that the distance between the source and the destination $D_{s,d}$ is maximized. Based on the SER performance obtained in Section 10.3, we can see that the performance of a cooperative UWB system is related not only to the power allocation but also to the location of the nodes. To maximize the distance $D_{s,d}$, it is obvious that the optimum relay location must be on the line joining the source and the destination, as shown in Fig. 10.4. This comes from the fact that if the relay is located in any location in a two-dimensional plane, its distances to both the source and the destination are always longer than their corresponding projections on the line joining the source and the destination. In this case the distance between the source and the destination can be written as a summation of the distance of the source–relay link and that of the relay–destination link (i.e., $D_{s,d} = D_{s,r} + D_{r,d}$). The question is: How far from the source should the relay be located and how much power should be applied at the source and the relay to maximize the distance $D_{s,d}$? To answer this question, we determine the distance $D_{s,r}$, the distance $D_{r,d}$, and the power ratio r such that the coverage range $D_{s,d}$ is maximized. We formulate an optimization as follows:

$$\max_{r, D_{s,r}, D_{r,d}} \quad D_{s,r} + D_{r,d} \tag{10.44}$$

$$\text{s.t.} \quad \begin{cases} \text{Performance: } P_e \le \varepsilon; \\ \text{Power: } rP \le P_{\max}, (1-r)P \le P_{\max}, 0 < r < 1. \end{cases}$$

To get some insights, we provide a solution to (10.44) without any constraint on the transmitter power level. In terms of the maximum power constraint, a solution similar to that at the end of Section 10.4.1 can be employed. As we show in Section 10.6, the solution to (10.44) with power constraint follows the same trend as that without power constraint. By applying the Lagrange multiplier method, solutions to (10.44) can be obtained by solving the first-order optimality conditions: $1 + \zeta \partial P_e / \partial D_{s,r} = 0$, $1 + \zeta \partial P_e / \partial D_{r,d} = 0$, $\partial P_e / \partial r = 0$, and $P_e - \varepsilon = 0$, where ζ is the Lagrange multiplier. Although the SER upper bound (10.15) and the asymptotic SER approximation (10.20) are simple, they are based on the assumption that all channel links are always available. Due to this assumption, the SERs (10.15) and (10.20) are not applicable for the problem in (10.44), in which two nodes can be located far away from each other. In what follows we are going to determine the optimum power allocation and the optimum distances based on the SER formulation (10.12) and the SER approximations (10.17) and (10.21).

We consider at first a UWB system with frequency spreading gain $g_F = 1$. According to the tight SER approximation (10.17) and the first-order optimality conditions, the optimum power allocation and distance must satisfy the necessary condition

$$\frac{A_1^2 r D_{s,r}^{-\nu-1}}{\left(1 + b\rho k r D_{s,r}^{-\nu}\right)^2} - \frac{A_2(1-r)D_{r,d}^{-\nu-1}}{\left(1 + b\rho k(1-r)D_{r,d}^{-\nu}\right)^2} = 0. \tag{10.45}$$

From (10.45) we can find the power ratio r as a function of the distances $D_{s,r}$ and $D_{r,d}$. Then, solving $\partial P_e / \partial r = 0$ and $P_e - \varepsilon = 0$ simultaneously, we obtain the optimum power ratio and distances $D_{s,r}$ and $D_{r,d}$. Similarly, the maximum coverage of a UWB system with frequency spreading gain $g_F = 2$ can be obtained as follows. By evaluating the first-order optimality conditions based on the approximate SER in (10.21), we obtain the necessary condition:

$$\frac{A_2^2 r D_{s,r}^{-\nu-1}\left[1 + 2b\rho k r\left(1 - B_{s,r}^2\right)D_{s,r}^{-\nu}\right]}{\left[1 + b\rho k r D_{s,r}^{-\nu} + b^2\rho^2 k^2 r^2\left(1 - B_{s,r}^2\right)D_{s,r}^{-2\nu}\right]^2}$$
$$- \frac{A_4(1-r)D_{r,d}^{-\nu-1}\left[1 + 2b\rho k(1-r)(1 - B_{r,d}^2)D_{r,d}^{-\nu}\right]}{\left[1 + b\rho k(1-r)D_{r,d}^{-\nu} + b^2\rho^2 k^2(1-r)^2\left(1 - B_{r,d}^2\right)D_{r,d}^{-2\nu}\right]^2} = 0. \tag{10.46}$$

Then the optimum power ratio r and optimum distances $D_{s,r}$ and $D_{r,d}$ can be determined by solving (10.46) together with $\partial P_e / \partial r = 0$ and $P_e - \varepsilon = 0$.

We also perform an exhaustive search to solve the optimization problem in (10.44) based on the SER formulation in (10.12). In Tables 10.3 and 10.4 we compare the optimum power allocation and the optimum distances obtained via exhaustive search and that obtained by solving the first-order optimality conditions. We consider a UWB multiband OFDM system with frequency spreading gains $g_F = 1$ in Table 10.3 and $g_F = 1$ in Table 10.4. Clearly, the analytical results closely match the results from an exhaustive search for all frequency spreading gains. Moreover, we can see that the optimum power allocation and the optimum relay location depends on the total power P/N_0. When P/N_0 is small, the maximum coverage can be achieved by putting the relay as far from the source as possible, and allocating almost all of the total transmitter power P at the source. However, when P/N_0 is high ($P/N_0 > 30$ dB), this is not the case. In such a scenario, putting the relay close to the middle and allocating about half of the power at the relay results in longer coverage than putting the relay farthest away from the source. We can explain these results intuitively as follows. At

TABLE 10.3 Power Allocation, Relay Location, and
Maximum Coverage of Cooperative UWB Multiband OFDM
Systems with Frequency Spreading Gain $g_F = 1$

P/N_0 (dB)	Exhaustive Search			Analytical Solution		
	r	$D_{s,r}$	$D_{s,d}$	r	$D_{s,r}$	$D_{s,d}$
25	0.86	13.00	14.06	0.88	13.74	14.87
30	0.86	23.12	25.01	0.83	23.37	25.70
35	0.55	15.53	33.82	0.58	15.12	33.98

TABLE 10.4 Power Allocation, Relay Location, and
Maximum Coverage of Cooperative UWB Multiband OFDM
Systems with Frequency Spreading Gain $g_F = 2$

P/N_0 (dB)	Exhaustive Search			Analytical Solution		
	r	$D_{s,r}$	$D_{s,d}$	r	$D_{s,r}$	$D_{s,d}$
25	0.89	17.11	19.14	0.88	17.31	19.79
30	0.85	30.24	35.81	0.84	30.17	35.46
35	0.52	13.21	43.87	0.54	13.27	43.92

a low SNR, the power transmitted is not large enough for the cooperation system to achieve a performance of diversity order 2. Therefore, the forwarding role of the relay is less important and we should use almost all of the power transmitted at the source. On the other hand, at high enough SNR, a diversity order of 2 can be achieved. In this case, the relay should be located in the middle to balance the channel quality of source–relay and relay–destination links.

10.5 IMPROVED COOPERATIVE UWB

The current multiband standard proposal [Bat03] allows several UWB devices to transmit at the same time using different subbands. However, in a short-range scenario, the number of UWB devices that transmit their information simultaneously tends to be smaller than the number of subbands available. Therefore, we can make use of the unoccupied subbands to improve the performance of cooperative UWB systems. The improved cooperative UWB strategy is as follows.

Time-domain spreading with a spreading factor of 2 is performed at the source. The improved cooperative UWB scheme comprises two phases, each corresponding to one OFDM symbol period. In phase 1 the source broadcasts its information to both destination and relay using one subband. In phase 2 the source repeats the information using another subband so as to gain the diversity from time spreading. At the same time, the relay forwards the source information using an unoccupied subband. The destination combines the signals received from the source in phases 1 and 2, and the signal from the relay in phase 2. Figure 10.5 is an example of an improved cooperative UWB system. In Fig. 10.5 the source and the relay are denoted

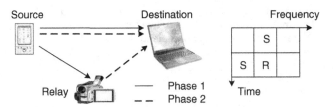

Figure 10.5 Improved cooperative UWB multiband OFDM scheme.

by S and R, respectively. It is worth noting that the improved cooperative UWB scheme is compatible with the current multiband standard proposal [Bat03], which allows multiuser transmission using different subbands. In addition, the cooperative UWB scheme yields the same data rate as the noncooperative scheme with the same spreading gain.

Similar to Section 10.2.1, we denote P_1 and P_2 as the power transmitted at the source in phases 1 and 2, respectively. The signals received from the direct link in phases 1 and 2 can be modeled as in (10.1). Let us denote P_3 as the power transmitted at the relay. Accordingly, the signal received from the relay link can be written as (10.4) by replacing $\tilde{P}_2$ with $\tilde{P}_3$. By the use of an MRC detector, the received signals $y_{s,d}^1$, $y_{s,d}^2$, and $y_{r,d}$ are optimally combined. The SNR of the MRC output can be expressed as

$$\eta = \frac{P_1}{N_0} \sum_{n \in \Phi_n} |H_{s,d}^1(n)|^2 + \frac{P_2}{N_0} \sum_{n \in \Phi_n} |H_{s,d}^2(n)|^2 + \frac{\tilde{P}_3}{N_0} \sum_{n \in \Phi_n} |H_{r,d}(n)|^2. \qquad (10.47)$$

Assuming ideal band hopping, the average SER of the improved cooperative UWB system is

$$P_e = \frac{1}{\pi^2} \int_0^{\pi - \pi/M} \mathcal{M}_{\eta_{s,d}}^2 \left(\frac{b}{\sin^2 \theta} \right) d\theta \int_0^{\pi - \pi/M} \mathcal{M}_{\eta_{s,r}} \left(\frac{b}{\sin^2 \theta} \right) d\theta$$

$$+ \frac{1}{\pi} \int_0^{\pi - \pi/M} \mathcal{M}_{\eta_{s,d}}^2 \left(\frac{b}{\sin^2 \theta} \right) \mathcal{M}_{\eta_{r,d}} \left(\frac{b}{\sin^2 \theta} \right) d\theta$$

$$\times \left[1 - \frac{1}{\pi} \int_0^{\pi - \pi/M} \mathcal{M}_{\eta_{s,r}} \left(\frac{b}{\sin^2 \theta} \right) d\theta \right]. \qquad (10.48)$$

Following the same procedures as in Section 10.3, we can approximate the SER in (10.48) as

$$P_e \approx \begin{cases} \dfrac{1}{(1 + b\rho_1 \sigma_{s,d}^2)(1 + b\rho_2 \sigma_{s,d}^2)} \left(\dfrac{A_1 A_2}{1 + b\rho_1 \sigma_{s,r}^2} + \dfrac{A_3}{1 + b\rho_3 \sigma_{r,d}^2} \right) & \text{if } g_F = 1; \\ & \qquad\qquad (10.49) \\ \dfrac{1}{g_{s,d}(\rho_1) g_{s,d}(\rho_2)} \left(\dfrac{A_2 A_4}{g_{s,r}(\rho_1)} + \dfrac{A_6}{g_{r,d}(\rho_3)} \right) & \text{if } g_F = 2. \\ & \qquad\qquad (10.50) \end{cases}$$

In (10.50) we denote

$$g_{x,y}(\rho_i) = 1 + b\rho_i\sigma_{x,y}^2 + b^2\rho_i^2\sigma_{x,y}^4\left(1 - B_{x,y}^2\right)$$

and $\rho_i = P_i/N_0$. If all channel links are available, the SER for the cooperative UWB system with frequency spreading gain $g_F = 1$ can be upper bounded at high SNR by

$$P_e \leq \frac{A_1 A_2}{b^3 \rho_1^2 \rho_2 \sigma_{s,d}^4 \sigma_{s,r}^2} + \frac{A_3}{b^3 \rho_1 \rho_2 \rho_3 \sigma_{s,d}^4 \sigma_{r,d}^2}. \tag{10.51}$$

With frequency spreading gain $g_F = 2$, the asymptotic SER performance can be approximated as

$$P_e \approx \frac{1}{b^4 \rho_1^2 \rho_2^2 \sigma_{s,d}^8\left(1 - B_{s,d}^2\right)} \left(\frac{A_2 A_4}{b^2 \rho_1^2 \sigma_{s,r}^4\left(1 - B_{s,r}^2\right)} + \frac{A_6}{b^2 \rho_3^2 \sigma_{r,d}^4\left(1 - B_{r,d}^2\right)}\right). \tag{10.52}$$

Suppose that the total power transmitted is $P_1 + P_2 + P_3 = P$, let $r_i = P_i/P$ for $i = 1, 2, 3$, and denote the power ratio of the transmitted power P_i over the total power P. The SER formulations in (10.51) and (10.52) can be written as

$$P_e \leq \left(\frac{b\left[\sigma_{s,d}^4 \sigma_{s,r}^2 \sigma_{r,d}^2 r_1^2 r_2\right]^{1/3}}{\left[A_1 A_2 \sigma_{r,d}^2 + A_3 \sigma_{s,r}^2 r_1/r_3\right]^{1/3}} \frac{P}{N_0}\right)^{-3} \quad \text{if } g_F = 1; \tag{10.53}$$

$$P_e \approx \left(\frac{b\left[r_1^4 r_2^2 \sigma_{s,d}^8 \sigma_{s,r}^4 \sigma_{r,d}^4\left(1 - B_{s,d}^2\right)^2\left(1 - B_{s,r}^2\right)\left(1 - B_{r,d}^2\right)\right]^{1/6}}{\left[A_2 A_4 \sigma_{r,d}^4\left(1 - B_{r,d}^2\right) + A_6 \sigma_{s,r}^4\left(1 - B_{s,r}^2\right)r_1^2/r_3^2\right]^{1/6}} \frac{P}{N_0}\right)^{-6} \quad \text{if } g_F = 2. \tag{10.54}$$

From (10.53) and (10.54) we can conclude that the improved cooperative UWB system provides an overall performance of diversity order $3g_F$. This confirms our expectation that the diversity order increases with the number of subbands used for transmission. Figure 10.6 depicts the SER performance of an improved cooperative UWB system as a function of P/N_0. We consider the UWB system with frequency spreading gains $g_F = 1$ and 2. The channel model parameters of each link are based on CM1. We can see that the theoretical formulations (10.48), (10.49), and (10.50) closely match the simulation curve. Moreover, the simple SER approximations (10.51) and (10.52) are tight at high SNR. Based on the SER formulations, we can determine the optimum power allocation for the improved cooperative UWB system as follows.

In the sequel we focus on minimizing the total power transmitted under the constraint on the error rate performance. Define $\mathbf{P} = [P_1 \ P_2 \ P_3]^T$ as a power allocation vector. Then the optimum power allocation can be determined by solving the problem in (10.31). As in Section 10.4.1, we first consider the problem (10.31) without the maximum power constraint to get some insight. By applying the Lagrange multiplier method and considering the first-order optimality conditions, we can show that the optimum power allocation vector $\mathbf{P}$ must satisfy the necessary conditions:

$$\frac{\partial P_e}{\partial P_1} = \frac{\partial P_e}{\partial P_2} = \frac{\partial P_e}{\partial P_3}. \tag{10.55}$$

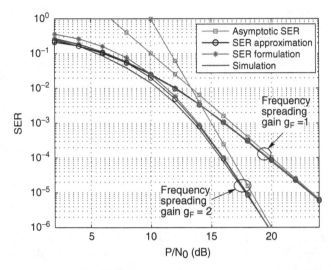

Figure 10.6 Comparison of the SER formulations and the simulation result for an improved cooperative UWB multiband OFDM system. We assume that $\sigma_{s,d}^2 = \sigma_{s,r}^2 = \sigma_{r,d}^2 = 1$ and $P_1 = P_2 = P_3 = P/3$.

Solving (10.55) and $P_e = \varepsilon$ simultaneously, we get the optimum power allocation **P**. Based on the tight SER approximation in (10.51), the asymptotic optimum power allocation for the improved cooperative UWB system with frequency spreading gain $g_F = 1$ can be determined as

$$P_1 = \frac{2rP}{3}, \quad P_2 = \frac{P}{3}, \quad \text{and} \quad P_3 = \frac{2(1-r)P}{3}, \tag{10.56}$$

where r is given in (10.35) and

$$P = \frac{N_0}{3b} \left(\frac{A_1 A_2 \sigma_{r,d}^2 + A_3 \sigma_{s,r}^2 r/(1-r)}{4\varepsilon r^2 \sigma_{s,d}^4 \sigma_{s,r}^2 \sigma_{r,d}^2} \right)^{1/3}. \tag{10.57}$$

The result in (10.56) reveals that the asymptotic optimum power allocation at the source in phase 2 does not depend on the channel link quality. That is, one-third of the total power transmitted should be allocated at the source in phase 2. Then the rest of the power is allocated at the relay and the source in phase 1 according to the channel quality of the source–relay and relay–destination links. Observe from (10.35) that r takes values between 1/2 and 1. This implies that more than one-third of P should be allocated at the source in phase 1, and less than one-third of P should be allocated at the relay. In case of frequency spreading gain $g_F = 2$, the asymptotic optimum power allocation is the same as (10.56), with r given in (10.37) and the total

TABLE 10.5   Comparisons Between Optimum Power Allocation Obtained Via Exhaustive Search and Analytical Results

Path Variance				Exhaustive Search			Solution in (10.56)		
$\sigma_{s,r}^2$	$\sigma_{s,r}^2$	$\sigma_{r,d}^2$	Gain, g_F	r_1	r_2	r_3	r_1	r_2	r_3
1	10	1	1	0.5367	0.3158	0.1476	0.5154	0.3333	0.1512
1	10	1	2	0.6175	0.2400	0.1425	0.5515	0.3333	0.1151
1	1	10	1	0.3530	0.3335	0.3135	0.3456	0.3333	0.3211
1	1	10	2	0.3374	0.3331	0.3294	0.3348	0.3333	0.3319

power P given by

$$
P = \frac{N_0}{3b} \left(\frac{A_2 A_4 \sigma_{r,d}^4 \left(1 - B_{r,d}^2\right) + A_6 \sigma_{s,r}^2 \left(1 - B_{s,r}^2\right) r / (1 - r)}{16 \varepsilon r^4 \sigma_{s,d}^8 \sigma_{s,r}^4 \sigma_{r,d}^4 \left(1 - B_{s,d}^2\right)\left(1 - B_{s,r}^2\right)\left(1 - B_{r,d}^2\right)} \right)^{1/6}. \tag{10.58}
$$

In Table 10.5 we compare the asymptotic optimum power allocation in (10.56) with the optimum power allocation obtained by exhaustive search based on the SER in (10.48). All channel links are based on CM1, and the target error rate performance is 5×10^{-2}. It is clear that the analytical solution in (10.56) agrees with the results from an exhaustive search. For a UWB system with the maximum power constraint, the power allocation can be determined by a procedure similar to that at the end of Section 10.4.1. Furthermore, the optimum power allocation that maximizes coverage can be obtained in a way similar to that in Section 10.4.2. We omit them here due to space limitations.

10.6 SIMULATION RESULTS

We perform computer simulations to compare the performance of cooperative UWB schemes and to validate the theoretical results derived in this paper. In all simulations we consider a UWB multiband OFDM system with 128 subcarriers and the subband bandwidth of 528 MHz. Each OFDM subcarrier is modulated using QPSK. We assume that the effect of intersymbol interference is mitigated by the use of a cyclic prefix. The propagation loss factor is $\nu = 2$ and the total multipath energy is modeled by $\sigma_{x,y}^2 = D_{x,y}^{-2}$. The channel model parameters follows those specified in the IEEE 802.15.3a standard [Foe03b]. In all simulations, the source is located at position $(0, 0)$.

In Fig. 10.7 we compare the average SER performances of UWB systems with different cooperation strategies. The locations of the relay and the destination are fixed at $(1 \text{ m}, 0)$ and $(2 \text{ m}, 0)$, respectively. All channel links are modeled by CM1. The total power transmitted is allocated equally. For a fair comparison, we present the SER curves as functions of P/N_0. From Fig. 10.7 we can see that both noncooperative and cooperative UWB systems achieve an overall performance of diversity order $2g_F$. For a frequency spreading gain $g_F = 1$, a cooperative UWB system outperforms a

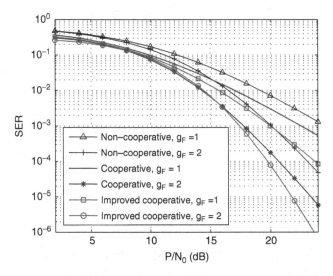

Figure 10.7 SER performance of UWB systems versus P/N_0.

noncooperative system with a SER performance of about 2 dB. This agrees with the analysis in (10.29), which shows that the performance gain of a cooperative UWB system compared with a noncooperative UWB system is $\xi = [(1 + A_1^2/A_2)\sigma_{s,d}^2]^{1/2} = 1.59$. For a frequency spreading gain $g_F = 2$, the performance of a cooperative system is about 2.5 dB better than that of a noncooperative system. This corresponds to the analysis in (10.30), in which the performance gain ξ can be calculated as $\xi = [(1 + A_2^2/A_4)\sigma_{s,d}^2]^{1/4} = 1.81$. Additionally, Fig. 10.7 illustrates that cooperative and improved cooperative UWB systems yield almost the same performance at low P/N_0. At high P/N_0, an improved cooperative UWB system provides performance of diversity order $3g_F$ and yields about a 2-dB performance improvement over a cooperative UWB system.

Figures 10.8 and 10.9 compare the total transmitter power of noncooperative and cooperative systems. We plot P/N_0 versus the destination location. In a cooperative system, the relay is located between the source and the destination (i.e., $D_{s,d} = D_{s,r}/2$). All channel links are modeled by CM4. The power transmitted is allocated such that overall transmitted power is minimized and the SER satisfies a performance requirement of 5×10^{-2}. In Fig. 10.8, we consider UWB systems without limitation on the power level transmitted. By increasing the frequency spreading gain from 1 to 2, the overall power transmitted can be reduced by 60%. With the same frequency spreading gain, the cooperative scheme achieves a 43% power saving over the noncooperative scheme. This is in-consistent with the analytical results in (10.41) and (10.42), in which the power ratio of cooperative and noncooperative schemes can be calculated as $P_{DF}/P_{NC} = 0.59$ for $g_F = 1$ and $P_{DF}/P_{NC} = 0.54$ for $g_F = 2$. Figure 10.8 also shows that using the improved cooperative UWB scheme can achieve up to a 52% power saving over the noncooperative scheme. In Fig. 10.9 we take into

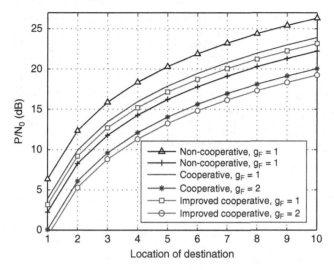

Figure 10.8 P/N_0 versus destination location.

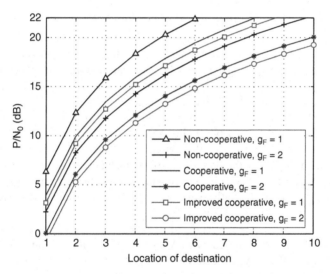

Figure 10.9 P/N_0 versus destination location for UWB systems with power limitation.

consideration the constraint on the transmitter power level and allocate the power based on the suboptimal solution provided in Section 10.4.1. The power limitation is set at $P_i/N_0 \leq 19$ dB. The tendencies observed in Fig. 10.9 are similar to those observed in Fig. 10.8. The improved cooperative scheme saves about 50% overall transmitted power when $g_F = 1$ and saves about 20% when $g_F = 2$.

Next, we study coverage of the UWB system under different cooperative strategies. All channel links are based on CM4. The SER performance requirement is fixed at 5 $\times 10^{-2}$. In Fig. 10.10 we plot the maximum distance between source and destination

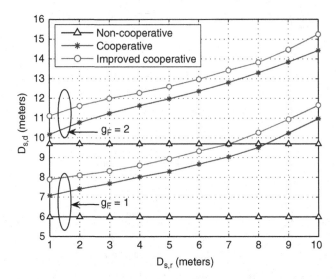

Figure 10.10 Distance between source and destination versus distance between source and relay.

versus the distance between source and relay for $P/N_0 = 22$ dB. We observe that by increasing the frequency spreading gain from 1 to 2, the noncooperative scheme increases the coverage by 60%, whereas the cooperative scheme increases the coverage by 40%. Moreover, coverage of the cooperative scheme increases as the relay is located farther away from the source. This agrees with our study in Section 10.4.2, which shows that at small P/N_0, the longer the distance between the source and the relay, the longer the distance between the source and the destination. For example, if the relay is located 1 m away from the source, the cooperative scheme increases the coverage by about 5%. On the other hand, if the distance between source and relay increases to 8 m, the cooperative scheme can increase the coverage by about 58% compared with the noncooperative scheme. With the improved cooperative scheme, the coverage can be increased by 70%.

In Fig. 10.11 we depict the coverage of a UWB system as a function of P/N_0. The transmitter power level is limited by $P_i/N_0 \leq 19$ dB. For a cooperative scheme, the relay location and the power allocation are designed such that the distance $D_{s,d}$ is maximized. We can see clearly from the figure that the coverage increases as P/N_0 increases. With the same P/N_0 and the same transmission data rate, the coverage of a UWB system can be increased up to 85% using the cooperative scheme, and it can be increased up to 100% using the improved cooperative scheme.

10.7 CHAPTER SUMMARY

In this chapter we enhance the performance of UWB systems by employing cooperative communications. We analyze SER performance and provide optimum power

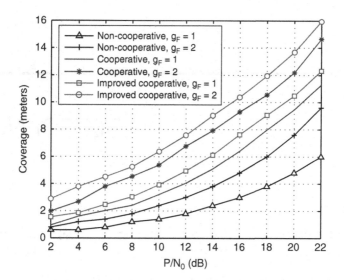

Figure 10.11 Maximum transmission range versus P/N_0.

allocation of cooperative UWB multiband OFDM systems with a decode-and-forward cooperative protocol. It turns out that both noncooperative and cooperative schemes achieve the same diversity order of twice the frequency spreading gain, which is independent of the clustering behavior of UWB channels. However, by taking advantage of the relay location and allocating the transmitted power properly, the cooperative UWB scheme can achieve performance superior to that of the noncooperative UWB scheme at the same data rate. We also further improve the performance of the cooperative UWB scheme by allowing the source and the relay nodes to retransmit the information simultaneously. With the objective of minimizing the overall power transmitted, we show by both theoretical and simulation results that the cooperative UWB multiband OFDM system can save up to 43% of the transmitter power. With the objective of maximizing coverage, both the optimum relay location and optimum power allocation depend on the SNR. At low SNR, the maximum coverage is achieved when the relay is located farthest away from the source and the source uses almost all of the transmitter power. On the other hand, at high SNR, the coverage is maximized when the relay is located in the middle between the source and the destination and approximately equal power is allocated at the source and the relay. Simulation results show that the cooperative UWB can increase the coverage range up to 85% compared with the noncooperative UWB scheme. By allowing both source and relay to retransmit data simultaneously, the improved cooperative UWB system achieves up to 52% power saving and up to 100% coverage extension.

REFERENCES

[Agr98]   D. Agrawal, V. Tarokh, A. Naguib,and N. Seshadri, "Space–time coded OFDM for high data-rate wireless communication over wideband channels," *Proc. IEEE Conf. Veh. Technol.*, vol. 3, pp. 2232–2236, 1998.

[Ala98]   S. Alamouti, "A simple transmit diversity technique for wireless communications," *IEEE J. Select. Areas Commun.*, vol. 16, no. 8, pp. 1451–1458, Oct. 1998.

[Alv03]   A. Alvarez et al., "New channel impulse response model for UWB indoor system simulations," *Proc. IEEE Semiannu. Veh. Technol. Conf.*, pp. 1–5, 2003.

[Bar00]   T. W. Barrett, "History of ultrawideband (UWB) radar and communications: pioneers and innovators," *Proc. Prog. Electromagn. Symp.*, Cambridge, MA, July 5–14, 2000.

[Bar04]   J. R. Barry, E. A. Lee, and D. G. Messerschmitt, *Digital Communication*, 3rd ed., Kluwer Academic Publishers, Norwood, MA, 2004.

[Bat03]   A. Batra et al., "Multi-band OFDM physical layer proposal for IEEE 802.15 task group 3a," IEEE P802.15-03/268r3, July 2003.

[Bat04]   A. Batra et al., "Design of a multiband OFDM system for realistic UWB channel environments," *IEEE Trans. Microwave Theory Tech.*, vol. 52, no. 9, pp. 2123–2138, Sept. 2004.

[Ber00]   H. L. Bertoni, *Radio Propagation for Modern Wireless Systems*, Prentice Hall, Upper Saddle River, NJ, 2000.

[Ber02]   I. Bergel, E. Fishler, and H. Messer, "Narrowband interference suppression in time-hopping impulse-radio systems," *Proc. IEEE Conf. Ultra Wideband Syst. Technol.*, pp. 303–307, 2002.

[Bla99]   N. Blaunstein, *Radio Propagation in Cellular Networks*, Artech House, Norwood, MA, 1999.

[Blu01]   R. Blum, Y. Li, J. Winters, and Q. Yan, "Improved space–time coding for MIMO-OFDM wireless communications," *IEEE Trans. Commun.*, vol. 49, pp. 1873–1878, Nov. 2001.

[Bol00]   H. Bölcskei and A. J. Paulraj, "Space–frequency coded broadband OFDM systems," *Proc. IEEE Wireless Commun. Network. Conf.*, pp. 1–6, Sept. 2000.

Ultra-Wideband Communications Systems: Multiband OFDM Approach, By W. Pam Siriwongpairat and K. J. Ray Liu
Copyright © 2008 John Wiley & Sons, Inc.

[Bou03] N. Boubaker and K. B. Letaief, "Ultra wideband DSSS for multiple access communications using antipodal signaling," *Proc. IEEE Int. Conf. Commun.*, vol. 3, pp. 11–15, May 2003.

[Bou04] N. Boubaker and K. B. Letaief, "Performance analysis of DS-UWB multiple access under imperfect power control," *IEEE Trans. Commun.*, vol. 52, no. 9, pp. 1459–1463, Sept. 2004.

[Boy04] J. Boyer, D. D. Falconer, and H. Yanikomeroglu, "Multihop diversity in wireless relaying channels," *IEEE Trans. Commun.*, vol. 52, no. 10, pp. 1820–1830, Oct. 2004.

[Bre01] M. Brehler and M. K. Varanasi, "Asymptotic error probability analysis of quadratic receivers in Rayleigh-fading channels with applications to a unified analysis of coherent and noncoherent space–time receivers," *IEEE Trans. Inf. Theory*, vol. 47, no. 6, pp. 2383–2399, Sept. 2001.

[Can03] C. M. Canadeo, M. A. Temple, R. O. Baldwin, and R. A. Raines, "Code selection for enhancing UWB multiple access communication performance using TH-PPM and DS-BPSK modulations," *Proc. IEEE Wireless Commun. Network. Conf.*, pp. 678–682, Mar. 2003.

[Cas02] D. Cassioli, M. Z. Win, F. Vatalaro, and A. F. Molisch, "Performance of low-complexity RAKE reception in a realistic UWB channel," *Proc. IEEE Int. Conf. Commun.*, vol. 2, pp. 763–767, 2002.

[Cha04] Y.-L. Chao and R. A. Scholtz, "Weighted correlation receivers for ultra-wideband transmitted reference systems," *Proc. IEEE Global Telecommun. Conf.*, vol. 1, pp. 66–70, Dec. 2004.

[Che03] Z. Chen, G. Zhu, J. Shen, and Y. Liu, "Differential space–time block codes from amicable orthogonal designs," *Proc. IEEE Wireless Commun. Network. Conf.*, pp. 768–772, Mar. 2003.

[Cho02] J. D. Choi and W. E. Stark, "Performance of ultra-wideband communications with suboptimal receivers in multipath channels," *IEEE J. Select. Areas Commun.*, vol. 20, no. 9, pp. 1754–1766, Dec. 2002.

[Cho04] C. Cho, H. Zhang, and M. Nakagawa, "A UWB repeater with a short relaying-delay for range extension," *Proc. IEEE Wireless Commun. Network. Conf.*, vol. 3, pp. 1436–1441, Mar. 2004.

[Cla01] K. L. Clarkson, W. Sweldens, and A. Zheng, "Fast multiple antenna differential decoding," *IEEE Trans. Commun.*, vol. 49, pp. 253–261, Feb. 2001.

[Cov91] T. M. Cover and J. A. Thomas, *Elements of Infomation Theory*, Wiley, New York, 1991.

[Cra98] R. J. Cramer, M. Z. Win, and R. A. Scholtz, "Evaluation of the multipath characteristics of the impulse radio channel," *Proc. 9th IEEE Int. Symp. Personal Indoor Mobile Radio Commun.*, New York, vol. 2, pp. 864–868, Sept. 1998.

[Cra02] R. J. Cramer, R. A. Scholtz, and M. Z. Win, "Evaluation of an ultra-wide-band propagation channel," *IEEE Trans. Antennas Propag.*, vol. 50, pp. 561–570, May 2002.

[Cuo02] F. Cuomo, C. Martello, A. Baiocchi, and F. Capriotti, "Radio resource sharing for ad hoc networking with UWB," *IEEE J. Select. Areas Commun.*, vol. 20, no. 9, pp. 1722–1732, Dec. 2002.

[Dig02]   S. N. Diggavi, N. Al-Dhahir, A. Stamoulis, and A. R. Calderbank, "Differential space–time coding for frequency-selective channels," *IEEE Commun. Lett.*, vol. 6, pp. 253–255, June 2002.

[Dis03]   Discrete Time Communications, "IEEE 802.15.3a 480Mbps wireless personal area networks: achieving a low complexity multi-band implementation," White Paper, Jan. 2003.

[Dur03]   G. Durgin, *Space–Time Wireless Channels*, Cambridge University Press, New York, 2003.

[Dur03a]   G. Durisi and S. Benedetto, "Performance evaluation and comparison of different modulation schemes for UWB multiaccess systems," *Proc. IEEE Int. Conf. Commun.*, vol. 3, pp. 2187–2191, May 2003.

[Dur03b]   G. Durisi and S. Benedetto, "Performance evaluation of TH-PPM UWB systems in the presence of multiuser interference," *IEEE Commun. Lett.*, vol. 7, no. 5, pp. 224–226, May 2003.

[ECM05]   ECMA-368, "High rate ultra wideband PHY and MAC standard," http://www. ecma-international.org/publications/files/ECMA-ST/ECMA-368.pdf, Dec. 2005.

[EIA92]   EIA/TIA IS-54, "Cellular system dual-mode mobile station–base station compatibility standard," EIA/TIA Technical Report IS-54, 1992.

[FCC02]   Federal Communications Commission, "Revision of Part 15 of the commission's rules regarding ultra-wideband transmission systems, first report and order," ET-Docket 98-153, FCC, Washington, DC, Feb. 2002.

[Fen04]   Z. Feng and T. Kaiser, "On channel capacity of multi-antenna UWB indoor wireless systems," *Proc. IEEE Int. Symp. Spread Spectrum Tech. Appl.*, Sydney, Australia, Aug. 30–Sept. 2, 2004.

[Fle06]   C. Fleming, "A tutorial on convolutional coding with Viterbi decoding," http:// home.netcom.com/chip.f/Viterbi.html, 2006.

[Foe02a]   J. R. Foerster, "The performance of a direct-sequence spread ultrawideband system in the presence of multipath, narrowband interference, and multiuser interference," *Proc. IEEE Conf. Ultra Wideband Syst. Technol.*, pp. 87–91, May 2002.

[Foe02b]   J. R. Foerster and Q. Li, "UWB channel modeling contribution from Intel," Technical Report P802.15 02/279SG3a, Intel Corporation, Hillsboro, OR, June 2002.

[Foe03a]   J. R. Foerster et al., "Intel CFP presentation for a UWB PHY," IEEE P802.15-03/109r1, Mar. 3, 2003.

[Foe03b]   J. R. Foerster et al., "Channel modeling sub-committee report final," IEEE 802.15-02/490, Nov. 18, 2003.

[Fon04]   R. J. Fontana, "Recent system applications of short-pulse ultra-wideband (UWB) technology," *IEEE Trans. Microwave Theory Tech.*, vol. 52, no. 9, pp. 2087–2104, Sept. 2004.

[Gan02]   G. Ganesan and P. Stoica, "Differential modulation using space–time block codes," *IEEE Signal Process. Lett.*, pp. 57–60, 2002.

[Gha02]   S. S. Ghassemzadeh et al., "A statistical path loss model for in-home UWB channels," *Proc. IEEE Conf. Ultra Wideband Syst. Technol.*, pp. 59–64, May 2002.

[Gon01]   Y. Gong and K. B. Letaief, "Space–frequency–time coded OFDM for broadband wireless communications," *Proc. IEEE Global Telecommun. Conf.*, vol. 1, pp. 519–523, Nov. 2001.

[Gue99]   J.-C. Guey, M. P. Fitz, M. R. Bell, and W.-Y. Kuo, "Signal design for transmitter diversity wireless communication systems over Rayleigh fading channels," *IEEE Trans. Commun.*, vol. 47, pp. 527–537, Apr. 1999.

[Hag88]   J. Hagenauer, "Rate-compatible punctured convolutional codes (RCPC codes) and their applications," *IEEE Trans. Commun.*, vol. 36, no. 4, pp. 389–400, Apr. 1988.

[Has93]   H. Hashemi, "Impulse response modeling of indoor radio propagation channels," *IEEE J. Select. Areas Commun.*, vol. 11, no. 7, pp. 967–978, Sept. 1993.

[Him05a] T. Himsoon, W. Su, and K. J. R. Liu, "Single-block differential transmit scheme for frequency selective MIMO-OFDM systems," *Proc. IEEE Wireless Commun. Network. Conf.*, vol. 1, pp. 532–537, Mar. 2005.

[Him05b] T. Himsoon, W. Su, and K. J. R. Liu, "Multiband differential modulation for UWB communication systems," *Proc. IEEE Global Telecommun. Conf.*, vol. 6, pp. 3789–3793, Dec. 2005.

[Him06]   T. Himsoon, W. Su, and K. J. R. Liu, "Single-block differential transmit scheme for broadband wireless MIMO-OFDM systems," *IEEE Trans. Signal Process.*, vol. 54, pp. 3305–3314, Sept. 2006.

[Ho02]   M. Ho, V. S. Somayazulu, J. Foerster, and S. Roy, "A differential detector for an ultra-wideband communications system," *Proc. IEEE Veh. Technol. Conf.*, vol. 4, no. 9, pp. 1896–1900, May 2002.

[Hoc99]   T. L. Marzetta and B. M. Hochwald, "Capacity of a mobile multiple-antenna communication link in Rayleigh flat fading," *IEEE Trans. Inf. Theory*, vol. 45, pp. 139–157, Jan. 1999.

[Hoc00]   B. M. Hochwald and T. L. Marzetta, "Unitary space–time modulation for multiple-antenna communication in Rayleigh flat fading," *IEEE Trans. Inf. Theory*, vol. 46, no. 2, pp. 543–564, Mar. 2000.

[Hoc01]   B. M. Hochwald and W. Sweldens, "Differential unitary space–time modulation," *IEEE Trans. Commun.*, vol. 48, pp. 2041–2052, Dec. 2000.

[Hor85]   R. A. Horn and C. R. Johnson, *Matrix Analysis*, Cambridge University Press, New York, 1985.

[Hug00]   B. L. Hughes, "Differential space–time modulation," *IEEE Trans. Inf. Theory*, vol. 46, pp. 2567–2578, Nov. 2000.

[Jaf01]   H. Jafarkhani and V. Tarokh, "Multiple transmit antenna differential detection from generalized orthogonal designs," *IEEE Trans. Inf. Theory*, pp. 2626–2631, Sept. 2001.

[Jan04]   M. Janani et al., "Coded cooperation in wireless communications: space–time transmission and iterative decoding," *IEEE Trans. Signal Process.*, vol. 52, pp. 362–370, Feb. 2004.

[Kai05]   T. Kaiser, Ed., *UWB Communications Systems: A Comprehensive Overview*, EURASIP Series on Signal Processing and Communications, Hindawi Publishing, New York, 2005.

[Kel97]   F. Kelly, "Charging and rate control for elastic traffic," *Eur. Trans. Telecommun.*, vol. 8, no. 1, pp. 33–37, Jan. 1997.

[Kel04]   H. Kellerer, U. Pferschy, and D. Pisinger, *Knapsack Problems*, Springer-Verlag, New York, 2004.

[Kum02]   N. A. Kumar and R. M. Buehrer, "Application of layered space–time processing to ultra wideband communication," *Proc. 45th Midwest Symp. Circuits Syst.*, vol. 3, pp. 597–600, Aug. 2002.

[Kun02]   J. Kunisch and J. Pamp, "Measurement results and modeling aspects for the UWB radio channels," *Proc. IEEE Conf. Ultra Wideband Syst. Technol.*, pp. 19–23, May 2002.

[Lai07]   H.-Q. Lai, W. P. Siriwongpairat, and K. J. R. Liu, "Performance analysis of MB-OFDM UWB system with imperfect synchronization and intersymbol interference," *Proc. 32nd IEEE Int. Conf. Acoust., Speech Signal Process.*, Apr. 2007.

[Leo94]   A. Leon-Garcia, *Probability and Random Processes for Electrical Engineering*, 2nd ed., Addison Wesley Longman, Boston, MA, 1994.

[Li03]   H. Li, "Differential space–time–frequency modulation over frequency-selective fading channels," *IEEE Commun. Lett.*, vol. 7, pp. 349–351, Aug. 2003.

[Liu02]   Z. Liu, Y. Xin, and G. Giannakis, "Space–time–frequency coded OFDM over frequency selective fading channels," *IEEE Trans. Signal Process.*, vol. 50, no. 10, pp. 2465–2476, Oct. 2002.

[Liu03]   Z. Liu, Y. Xin, and G. B. Giannakis, "Linear constellation precoding for OFDM with maximum multipath diversity and coding gains," *IEEE Trans. Commun.*, vol. 51, no. 3, pp. 416–427, Mar. 2003.

[Ma03]   Q. Ma, C. Tepedelenlioğlu, and Z. Liu, "Full diversity block diagonal codes for differential space–time–frequency coded OFDM," *Proc. IEEE Global Telecommun. Conf.*, vol. 2, pp. 868–872, Dec. 1–5, 2003.

[Mat92]   A. M. Mathai and S. B. Provost, *Quadratic Forms in Random Variables: Theory and Applications*, Marcel Dekker, New York, 1992.

[Mol02]   A. F. Molisch, M. Z. Win, and J. H. Winters, "Space–time–frequency (STF) coding for MIMO-OFDM systems," *IEEE Commun. Lett.*, vol. 6, no. 9, pp. 370–372, Sept. 2002.

[Mol03]   A. F. Molisch, J. R. Foerster, and M. Pendergrass, "Channel models for ultrawideband personal area networks," *IEEE Wireless Commun.*, vol. 10, no. 6, pp. 14–21, Dec. 2003.

[Nak60]   M. Nakagami, "The *m*-distribution: a general formula of intensity distribution of rapid fading," in *Statistical Methods in Radio Wave Propagation*, W. G. Hoffman, Ed., Pergamon Press, Oxford, 1960.

[Nak04]   Y. Nakache et al., "Low-complexity ultrawideband transceiver with compatibility to multiband-OFDM," Technical Report, Mitsubishi Electronic Research Laboratory, www.merl.com/reports/docs/TR2004-051.pdf, 2004.

[Opp04]   I. Oppermann et al., "UWB wireless sensor networks: UWEN—a practical example," *IEEE Commun. Mag.*, vol. 42, no. 12, pp. 27–32, Dec. 2004.

[Pen02]   M. Pendergrass and W. C. Beeler, "Empirically based statistical ultra-wideband (UWB) channel model," Technical Report P802.15 02/240SG3a, Time Domain Corporation, Huntsville, AL, June 2002.

[Por03]   D. Porcino and W. Hirt, "Ultra-wideband radio technology: potential and challenges ahead," *IEEE Commun. Mag.*, vol. 41, no. 7, pp. 66–74, July 2003.

[Pro01]   J. G. Proakis, *Digital Communications*, 4th ed., McGraw-Hill, New York, 2001.

[Rad04] B. Radunovic and J.-Y. Le Boudec, "Optimal power control, scheduling, and routing in UWB networks," *IEEE J. Select. Areas Commun.*, vol. 22, no. 7, pp. 1252–1270, Sept. 2004.

[Ram98] F. Ramirez-Mireles and R. A. Scholtz, "Multiple-access performance limits with time hopping and pulse position modulation," *Proc. IEEE Military Commun. Conf.*, pp. 529–533, Oct. 1998.

[Rob03] R. Roberts, "XtremeSpectrum CFP document," IEEE P802.15-03/154r1, Mar. 3, 2003.

[Ros63] G. F. Ross, "The transient analysis of multiple beam feed networks for array systems," Ph.D. dissertation, Polytechnic Institute of Brooklyn, Brooklyn, NY, 1963.

[Ros73] G. F. Ross, "Transmission and reception system for generating and receiving base-band duration pulse signals for short base-band pulse communication system," U.S. patent 3,728,632, Apr. 17, 1973.

[Sab03] E. Saberinia and A. H. Tewfik, "Pulsed and non-pulsed OFDM ultra wideband wireless personal area networks," *Proc. IEEE Conf. Ultra Wideband Syst. Technol.*, pp. 275–279, Nov. 2003.

[Sab04] E. Saberinia, J. Tang, A. H. Tewfik, and K. K. Parhi, "Design and implementation of multi-band pulsed-OFDM system for wireless personal area networks," *Proc. IEEE Int. Conf. Commun.*, vol. 2, pp. 862–866, June 20–24, 2004.

[Sad05] A. K. Sadek, W. Su, and K. J. R. Liu, "Performance analysis for multi-node decode-and-forward relaying in cooperative wireless networks," *Proc. IEEE Int. Conf. Acoust. Speech Signal Process.*, vol. 3, pp. 521–524, Mar. 2005.

[Sal87] A. A. M. Saleh and R. A. Valenzuela, "A statistical model for indoor multi-path propagation," *IEEE J. Select. Areas Commun.*, vol. 5, no. 2, pp. 128–137, Feb. 1987.

[Sch93] R. A. Scholtz, "Multiple access with time-hopping impulse modulation," *Proc. MILCOM Conf.*, Boston, MA, pp. 447–450, Oct. 1993.

[Sen03a] A. Sendonaris, E. Erkip, and B. Aazhang, "User cooperation diversity, part I: system description," *IEEE Trans. Commun.*, vol. 51, no. 11, pp. 1927–1938, Nov. 2003.

[Sen03b] A. Sendonaris, E. Erkip, and B. Aazhang, "User cooperation diversity, part II: implementation aspects and performance analysis," *IEEE Trans. Commun.*, vol. 51, no. 11, pp. 1939–1948, Nov. 2003.

[Sim00] M. K. Simon and M.-S. Alouini, *Digital Communication over Fading Channels: A Unified Approach to Performance Analysis*, Wiley, New York, 2000.

[Sim04] M. K. Simon and M.-S. Alouini, *Digital Communication over Fading Channels*, 2nd ed., Wiley, Hoboken, NJ, 2004.

[Sir04] W. P. Siriwongpairat, M. Olfat, and K. J. R. Liu, "On the performance evaluation of TH and DS UWB MIMO systems," *Proc. IEEE Wireless Commun. Network. Conf.*, vol. 3, pp. 1800–1805, Mar. 2004.

[Sir05a] W. P. Siriwongpairat, M. Olfat, and K. J. R. Liu, "Performance analysis and comparison of time hopping and direct sequence UWB-MIMO systems," *EURASIP J. Appl. Signal Process.*, Special Issue on 'UWB: State of the Art,' vol. 2005, no. 3, pp. 328–345, Mar. 2005.

[Sir05b]   W. P. Siriwongpairat, Z. Han, and K. J. R. Liu, "Energy-efficient resource allocation for multiband UWB communication systems," *Proc. IEEE Wireless Commun. Network. Conf.*, vol. 2, pp. 813–818, Mar. 2005.

[Sir05c]   W. P. Siriwongpairat, W. Su, M. Olfat, and K. J. R. Liu, "Space–time–frequency coded multiband UWB communication systems," *Proc. IEEE Wireless Commun. Network. Conf.*, vol. 1, pp. 426–431, Mar. 2005.

[Sir05d]   W. P. Siriwongpairat, W. Su, and K. J. R. Liu, "Characterizing performance of multiband UWB systems using Poisson cluster arriving fading paths," *Proc. 6th IEEE Workshop Signal Process. Adv. Wireless Commun.*, pp. 246–250, June 2005.

[Sir06a]   W. P. Siriwongpairat, W. Su, M. Olfat, and K. J. R. Liu, "Multiband-OFDM MIMO coding framework for UWB communication systems," *IEEE Trans. Signal Process.*, vol. 54, no. 1, pp. 214–224, Jan. 2006.

[Sir06b]   W. P. Siriwongpairat, W. Su, and K. J. R. Liu, "Performance characterization of multiband UWB communication systems using Poisson cluster arriving fading paths," *IEEE J. Select. Areas in Commun.*, Issue on Ultra Wideband Wireless Communications: Theory and Applications, vol. 24, no. 4, pp. 745–751, Apr. 2006.

[Sir06c]   W. P. Siriwongpairat, W. Su, Z. Han, and K. J. R. Liu, "Employing cooperative diversity for performance enhancement in UWB communication systems," *Proc. IEEE Wireless Commun. Network. Conf.*, vol. 4, pp. 1854–1859, Apr. 2006.

[Sir07]   W. P. Siriwongpairat, Z. Han, and K. J. R. Liu, "Power controlled channel allocation for multiuser multiband UWB systems," *IEEE Trans. Wireless Commun.*, vol. 6, no. 2, pp. 583–592, Feb. 2007.

[Som02]   V. S. Somayazulu, "Multiple access performance in UWB systems using time hopping vs. direct sequence spreading," *Proc. IEEE Wireless Commun. Network. Conf.*, vol. 2, pp. 522–525, Mar. 2002.

[Str01]   A. Street, L. Lukama, and D. Edwards, "Use of VNAs for wideband propagation measurements," *IEE Proc. Commun.*, vol. 148, no. 6, pp. 411–415, Dec. 2001.

[Stu00]   G. L. Stüber, *Principles of Mobile Communications*, 2nd ed., Kluwer Academic Publishers, Norwell, MA, 2000.

[Su03]   W. Su, Z. Safar, M. Olfat, and K. J. R. Liu, "Obtaining full-diversity space–frequency codes from space–time codes via mapping," *IEEE Trans. Signal Process.*, vol. 51, no. 11, pp. 2905–2916, Nov. 2003.

[Su04a]   W. Su, Z. Safar, and K. J. R. Liu, "Systematic design of space–frequency codes with full rate and full diversity," *Proc. IEEE Wireless Commun. Network. Conf.*, vol. 3, pp. 1436–1441, Mar. 2004.

[Su04b]   W. Su and K. J. R. Liu, "Differential space–frequency modulation for MIMO-OFDM systems via a smooth logical channel," presented at IEEE Global Telecommun. Conf., Dec. 2004.

[Su05a]   W. Su, Z. Safar, and K. J. R. Liu, "Full-rate full-diversity space–frequency codes with optimum coding advantage," *IEEE Trans. Inf. Theory*, vol. 51, no. 1, pp. 229–249, Jan. 2005.

[Su05b]   W. Su, A. K. Sadek, and K. J. R. Liu, "SER performance analysis and optimum power allocation for decode-and-forward cooperation protocol in wireless networks," *Proc. IEEE Wireless Commun. Network. Conf.*, vol. 2, pp. 984–989, Mar. 2005.

[Su05c]   W. Su, Z. Safar, and K. J. R. Liu, "Towards maximum achievable diversity in space, time and frequency: performance analysis and code design," *IEEE Trans. Wireless Commun.*, vol. 4, no. 4, pp. 1847–1857, July 2005.

[Tao01]   M. Tao and R. S. Cheng, "Differential space–time block codes," *Proc. IEEE Global Telecommun. Conf.*, pp. 1098–1102, Nov. 2001.

[Tar98]   V. Tarokh, N. Seshadri, and A. R. Calderbank, "Space–time codes for high data rate wireless communication: performance criterion and code construction," *IEEE Trans. Inf. Theory*, vol. 44, no. 2, pp. 744–765, Mar. 1998.

[Tar99]   V. Tarokh, H. Jafarkhani, and A. R. Calderbank, "Space–time block codes from orthogonal designs," *IEEE Trans. on Inf. Theory*, vol. 45, no. 5, pp. 1456–1467, July 1999.

[Tar00]   V. Tarokh and H. Jafarkhani, "A differential detection scheme for transmit diversity," *IEEE J. Select. Areas Commun.*, vol. 18, pp. 1169–1174, July 2000.

[Tar03]   S. S. Ghassemzadeh, L. J. Greenstein, T. Sveinsson, and V. Tarokh, "A multipath intensity profile model for residential environments," *Proc. IEEE Wireless Commun. Network. Conf.*, vol. 1, pp. 150–155, Mar. 2003.

[TG3]   IEEE 802.15WPAN High Rate Task Group 3 (TG3). www.ieee802.org/15/pub/TG3.html.

[TG3a]   IEEE 802.15WPAN High Rate Alternative PHY Task Group 3a (TG3a). www.ieee802.org/15/pub/TG3a.html.

[TG4a]   IEEE 802.15 WPAN Low Rate Alternative PHY Task Group 4a (TG4a). www.ieee802.org/15/pub/TG4a.html.

[Wan02]   J. Wang and K. Yao, "Differential unitary space–time–frequency coding for MIMO OFDM systems," *Conf. Rec. 36th Asilomar Conf. Signals Syst. Comput.*, vol. 2, pp. 1867–1871, Nov. 3–6, 2002.

[Wei03]   M. Weisenhorn and W. Hirt, "Performance of binary antipodal signaling over the indoor UWB MIMO channel," *Proc. IEEE Int. Conf. Commun.*, vol. 4, pp. 2872–2878, May 2003.

[Wel01]   M. L. Welborn, "System considerations for ultra-wideband wireless networks," *Proc. IEEE Radio Wireless Conf.*, pp. 5–8, Aug. 2001.

[Win98]   M. Z. Win and R. A. Scholtz, "Impulse radio: how it works," *IEEE Commun. Lett.*, vol. 2, no. 2, pp. 36–38, Feb. 1998.

[Win00]   M. Z. Win and R. A. Scholtz, "Ultra-wide bandwidth time-hopping spread-spectrum impulse radio for wireless multiple-access communications," *IEEE Trans. Commun.*, vol. 48, no. 4, pp. 679–691, Apr. 2000.

[Win02]   D. Cassioli, M. Z. Win, and A. F. Molisch, "The ultra-wide bandwidth indoor channel: from statistical model to simulations," *IEEE J. Select. Areas Commun.*, vol. 20, pp. 1247–1257, Aug. 2002.

[Won99]   C. Y. Wong, R. S. Cheng, K. B. Letaief, and R. D. Murch, "Multiuser OFDM with adaptive subcarrier, bit, and power allocation," *IEEE J. Select. Areas Commun.*, vol. 17, no. 10, pp. 1747–1758, Oct. 1999.

[Wor03]   J. N. Laneman and G. W. Wornell, "Distributed space–time coded protocols for exploiting cooperative diversity in wireless networks," *IEEE Trans. Inf. Theory*, vol. 49, pp. 2415–2525, Oct. 2003.

[Wor04]   J. N. Laneman, D. N. C. Tse, and G. W. Wornell, "Cooperative diversity in wireless networks: efficient protocols and outage behavior," *IEEE Trans. Inf. Theory*, vol. 50, no. 12, pp. 3062–3080, Dec. 2004.

[Yan02]   L. Yang and G. B. Giannakis, "Space–time coding for impulse radio," *Proc. IEEE Conf. Ultra Wideband Syst. Technol.*, pp. 235–239, May 2002.

[Zha03]   J. Zhang, R. A. Kennedy, and T. D. Abhayapala, "Performance of RAKE reception for ultra wideband signals in a lognormal fading channel," presented at Int. Workshop on UWB Systems, Oulo, Finland, June 2003.

[Zhu02]   F. Zhu, Z. Wu, and C. R. Nassar, "Generalized fading channel model with applications to UWB," *Proc. IEEE Conf. Ultra Wideband Syst. Technol.*, pp. 13–17, 2002.

[Zhu03]   W. Zhuang, X. Shen, and Q. Bi, "Ultra-wideband wireless communications," *Wireless Commun. Mobile Comput.*, Special Issue on Ultra-Broadband Wireless Communications for the Future, Invited Paper, vol. 3, no. 6, pp. 663–685, 2003.

INDEX

Ultra-Wideband Communications Systems: Multiband OFDM Approach, By W. Pam Siriwongpairat and K. J. Ray Liu
Copyright © 2008 John Wiley & Sons, Inc.

WILEY SERIES IN TELECOMMUNICATIONS AND SIGNAL PROCESSING

John G. Proakis, Editor
Northeastern University

Printed in the United States
By Bookmasters